Towards Excellence in
Engineering Education

Towards Excellence in Engineering Education

Special Issue Editor

Khmaies Ouahada

MDPI • Basel • Beijing • Wuhan • Barcelona • Belgrade

MDPI

Special Issue Editor
Khmaies Ouahada
University of Johannesburg
South Africa

Editorial Office
MDPI
St. Alban-Anlage 66
4052 Basel, Switzerland

This is a reprint of articles from the Special Issue published online in the open access journal *Education Sciences* (ISSN 2227-7102) from 2018 to 2019 (available at: https://www.mdpi.com/journal/education/special issues/Towards Excellence in Engineering Education)

       –   –   – –   –

For citation purposes, cite each article independently as indicated on the article page online and as indicated below:

LastName, A.A.; LastName, B.B.; LastName, C.C. Article Title. *Journal Name* **Year**, *Article Number*, Page Range.

ISBN 978-3-03921-251-4 (Pbk)
ISBN 978-3-03921-252-1 (PDF)

Contents

About the Special Issue Editor

Khmaies Ouahada (Prof) is a full professor and head of the Department of Electrical and Electronic Engineering Science, Faculty of Engineering and the Built Environment, at the University of Johannesburg, Johannesburg, South Africa.

He holds a doctorate in information theory and telecommunications. His research interests are information theory, coding techniques, power-line communications, visible light communications, smart grid, energy demand management, renewable energy, wireless sensor networks, wireless communications, reverse engineering, and engineering education.

Prof Khmaies Ouahada was awarded the 2016 University of Johannesburg Vice-Chancellor's Distinguished Award for Teaching Excellence. He is the founder and chairman of the Centre for Smart Systems (CSS) research group in the department. He serves as an Assistant Editor for the IEEE Access journal, USA, and is a member of the Editorial Board of the Multidisciplinary Digital Publishing Institute (MDPI) journals Sustainability and Information and Digital Communications and Networks, Elsevier. He was the Guest Editor of the Special Issue "Engineering Education" in MDPI's Education Sciences and "Energies and Coding and Modulation Techniques" in MDPI's journal Information.

Prof Khmaies Ouahada is a member of the UJ Academy for Teaching Excellence at the University of Johannesburg and the chair of the "Smart Communications Systems" track for the International Symposium on Networks, Computers and Communications (ISNCC) conferences. He was the publications chair for the 2013 International Symposium on Power Line Communications and its Applications (ISPLC), held at the University of Johannesburg, South Africa.

Prof Khmaies Ouahada is a member of the South African 5G Forum representing the University of Johannesburg and the founder and chairperson of the Academics Beyond Academia (ABA) committee for community engagement at the Faculty of Engineering and the Built Environment.

Preface to "Towards Excellence in Engineering Education"

Engineers play different contextual roles in industry and academia, not only by teaching but by mentoring, supervising and training students. Engineers are educators who are expected to provide their students with authentic learning experiences that are relevant to contemporary concerns, and to produce engineers who are responsible, insightful, work independently, have favorable problem-solving skills, and can apply and adapt their knowledge to new and unexpected situations.

This book from Education Sciences focuses on important issues in engineering education. In this Special Issue entitled "Towards Excellence in Engineering Education" we invite educators and researchers from engineering universities to discuss and share their experiences. What makes engineering education different to other educational disciplines? What are the challenges faced by engineering education and how should the educational system and curriculum be designed to cope with the high-speed development of technology?

This book highlights 11 papers that cover a diverse range of topics of engineering education, mainly focusing on lecturers' personal experiences in engineering education shared through teaching portfolios, assessment styles and teaching methods. E-learning in engineering education is also covered in this book as many lecturers in the engineering field use technology to select, design, deliver, administer, facilitate, and support learning. Examples include computer-based, web-based, and mobile learning.

The book covers curriculum in engineering education that offers rigorous analysis of theoretical principles as well as intensive hands-on experience. The engineering curriculum can be divided into three branches, namely engineering science, systems, and design and professional practice.

Here, the authors present some of their contributions and the experiences they used to assess engineering students. The academics share the modern teaching methods they use in engineering education, for example, active classrooms, flipped classrooms, problem-based learning and many more that are suitable to the nature of engineering disciplines.

This book highlights engineering education for community engagement. EPICS (engineering projects in community service) is an educational program that combines ideas surrounding teaching and learning with the community. Teams of students participate with local and global community organizations to address human, community, and environmental needs.

Khmaies Ouahada
Special Issue Editor

education sciences

MDPI

Article

Course Evaluation for Low Pass Rate Improvement in Engineering Education

Khmaies Ouahada

Department of Electrical and Electronic Engineering Science, University of Johannesburg, Johannesburg 2092, South Africa; kouahada@uj.ac.za; Tel.: +27-11-559-2213

Received: 3 April 2019; Accepted: 20 May 2019; Published: 2 June 2019

Abstract: A course evaluation is a process that includes evaluations of lecturers' teaching performances and their course material moderations. These two procedures are usually implemented, whether officially by the faculty of engineering or by lecturers' own initiatives, to help identify lecturers' strengths and weaknesses and the ways forward to improve their performances and their qualities of teaching. This paper presents different ways of implementing these two criteria from students' and professionals' perspectives. Official questionnaires from the faculty of engineering, personal questionnaires using Google surveys, Moodle and special designed forms have been used for moderation and evaluations. The process of evaluation is the core of a feedback procedure followed by universities in order for them to monitor the teaching quality of their staff. Satisfactory results show that such a process can improve the lecturers' teaching performances, courses material quality, students' satisfaction and performances, and finally the pass rate of the class.

Keywords: education; engineering; evaluation; survey; feedback; moderation; pass rate; module

1. Introduction

Engineers play different contextual roles in industry and academia. In the latter, they not only teach students, but are also regarded as mentors and expected to extend open door policies to their students.

The teaching practice should be informed by the lecturers' working environment, namely the Faculty of Engineering, and their professional statuses as educators in the 21st century. Lecturers should be motivated by the importance of providing students with authentic learning experiences which are relevant to contemporary concerns, and place high value on developing responsible engineers who are insightful, can work independently, have good problem solving skills, and can apply and adapt their knowledge to unexpected and new situations.

It is known that the evaluation [1] of lecturers' teaching quality is usually conducted for two reasons; the improvement of practice, since more experience can be built up from the received feedbacks; and the faculty of engineering promotion, which is subject to the university policy for staff promotion as proof of teaching evaluation should be required. Also, the evaluation can be conducted by students or professionals—either colleagues or visiting experts appointed by the faculty of engineering. In the case of students' evaluations, the most important benefit lecturers can gain is feedback to help them refine their courses and teaching practices to provide students with better learning experiences [2–4].

The question here is: how important is the evaluation in the improvement of the low passing rate of the offered courses [5]. Although the evaluation's impact on the students' success cannot be demonstrated clearly, it can assist lecturers to improve their teaching style and upgrade their course materials, which usually have direct impact on students' performances. It can be seen that evaluation processes can accurately identify the lecturers' strengths as well as areas in which they need to improve.

Usually, a course can be defined into four major parts; the prescribed textbooks, lecture notes, tutorials, and the practicals. Another important item can be useful to make the course easier to follow and more comprehensive is the study guide. Study guides are very important in the organization of the course. They are the road map and can be seen as contracts between lecturers and their students.

Prescribed textbooks are usually the official books for the course. Lecturers, after obtaining faculty approval, list these books as essential for reading and reference. These books will help students to focus better and supplement lecture notes, creating better chances of student success in the course.

Lecturers are advised to avoid recommended classic textbooks due to their outdated contents and applications. These books were basically designed for students who had very limited access to computers and digital information. Modern engineering textbooks should be user-friendly, with new and modern applications inspired from the modern engineering world. Tutorials in these textbooks ought to be designed to solve real-life problems using pedagogical approaches that help students understand the course through their own studies and revisions.

In order to give students greater variety and inspire them to think out-of-the-box and not to rely on what is prescribed to them, lecturers could also recommend textbooks written by different authors and prescribed by other universities. The recommended textbooks will not replace lecturer-prescribed textbooks but give a chance to students to expand their horizons through exposure to something different. Usually, these textbooks are provided to students in the form of e-books for no extra expense.

Good design and presentation of lecture notes or course slides, despite the brevity of the latter, play an important role in making lectures very easy to understand and comprehensive. Succinct and well-summarized lecture notes help lecturers cope with the limited time allocated per session. Lecture notes facilitate revision for students.

Lecture notes should be designed and prepared in an attractive manner and the layout of the slides helps students psychologically follow the content when lecturers are busy presenting. The slides should have an easy logical flow and should be judiciously interspersed with some proverbs, photographs, or cartoons in line with the context of the lecture.

This added entertainment aspect could also include information related to the content of the course gleaned from famous researchers or well-known scientists and gives quality to the design of the slides. The extra information provides inspiration to students and assists them to see the course from a real-word perspective.

There is a famous quote from Socrates: I cannot teach anybody anything, I can only make them think. This quote describes the philosophy behind tutorials at engineering faculties. The main purpose of a tutorial is to give students a chance to develop their individual capacities to think deeply about engineering problems and thereby build their confidence.

Tutorials also encourage teamwork among students when they meet in small groups and discuss specific topics related to the subject matter of the course. Engineering tutorials which involve group work are appealing since they provide opportunities for students to practice and develop collaborative skills [6]. In a tutorial class, the lecturer will encourage interaction and participation in the discussion.

As tutorials are very important in mediating the course by helping students grasp the unclear concepts, the lecturer should link the problems given in tutorials to the theory in the lecture notes. Solutions to the most important problems should also be made available to students to enhance their understanding of the lecture material.

The overall goal of engineering education is to prepare students to practice engineering [7]. Therefore, practicals in engineering education play an important role in developing skills that will assist students to be ready for the professional engineering environment.

Engineering faculties consider laboratories as an essential part of undergraduate programs. Laboratory work is an established part of courses in engineering education that intends to produce skilled and highly competent engineers for industry. This enables students to integrate easily and quickly into industry.

Practicals enable students to link theoretical concepts learned in class to real-life applications. For example, practicals designed for the Signals and Systems and Telecommunications course constitute either software programming projects or hardware build projects or a combination of both.

The architecture of the proposed procedure for improvement of low class pass rate is depicted in Figure 1. Feedback is the core of the system. The teaching and course material evaluations are means to provide feedback about the course quality and thus help lecturers to improve and upgrade what is necessary in order to provide students with a better educational environment which will lead to better class pass rates.

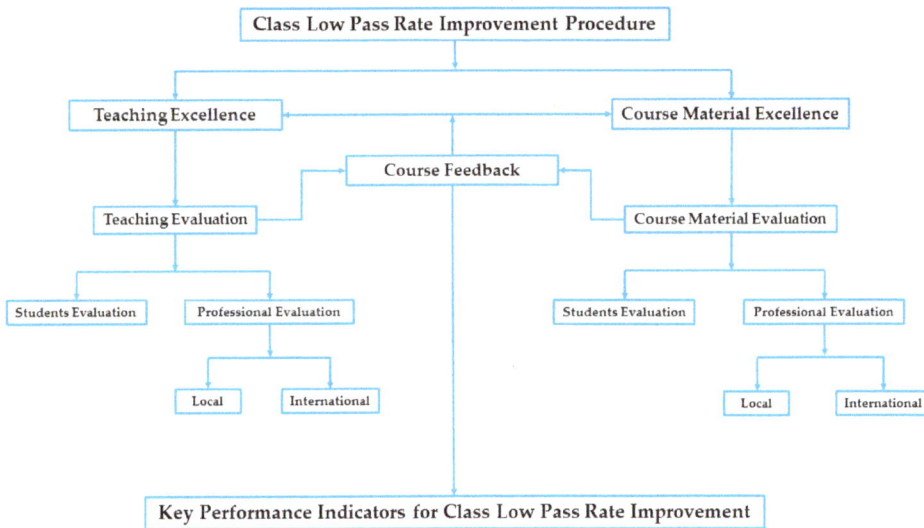

Figure 1. System architecture of the proposed procedure.

Google's survey service simplifies communication between lecturers and students and summarizes the collected data based on students' opinions and views. Figure 2 is one of the surveys [8–10] administered to students regarding teaching expertise and course material. Lecturers also use Moodle [11,12] to reach a decision when it comes to meeting and test dates. This help create sort of a cloud-community that maintains open and flexible access and communication between lecturers and students even beyond physical universities.

The paper is organized as follows. In Section 2, different evaluation processes are presented. Section 3 covers the process of moderation of the course material from the students' perspective and professional academics' perspective. Finally, a conclusion summarizing the achievements which led to the improvement of the class pass rate is presented in Section 4.

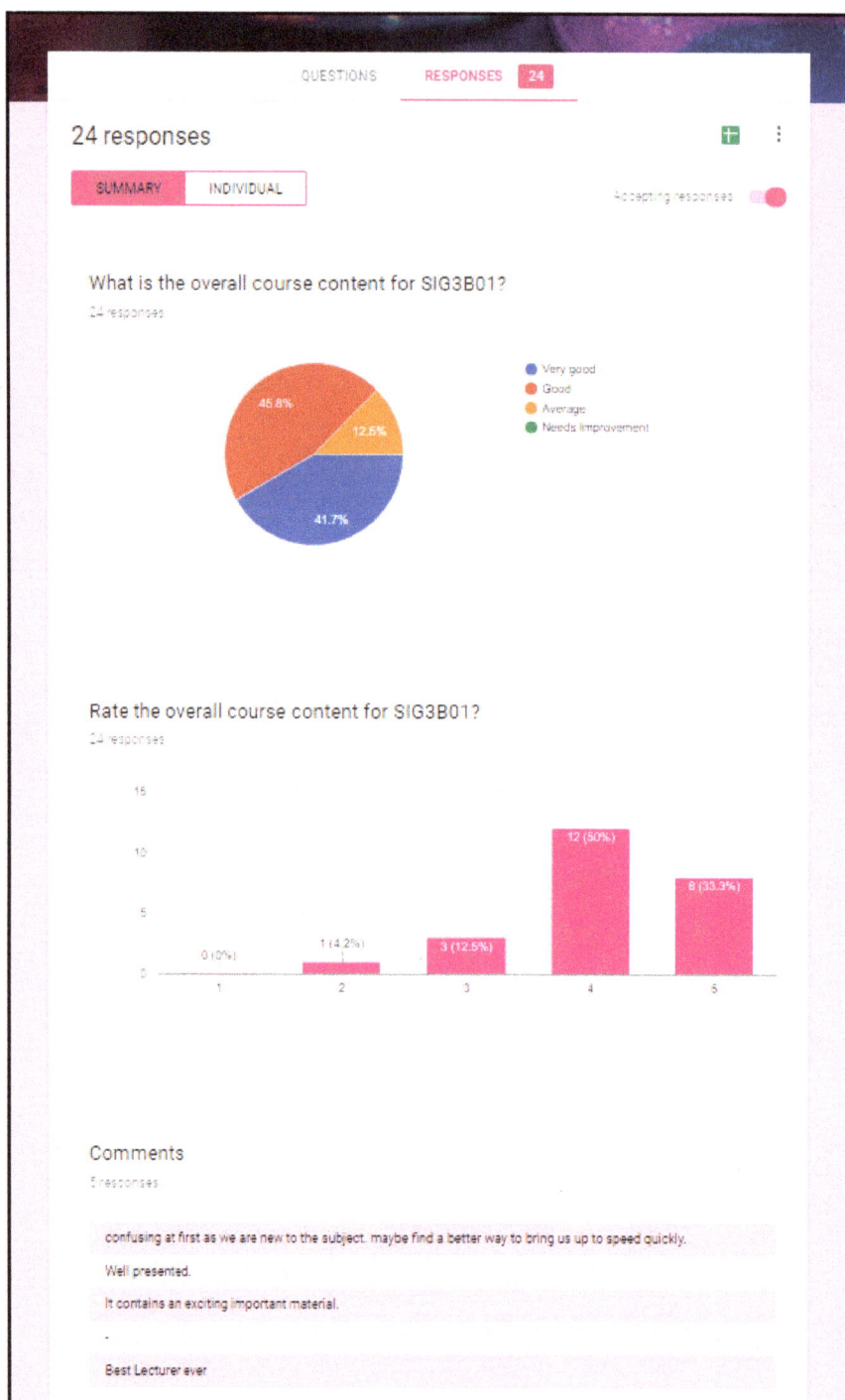

Figure 2. Google Survey for teaching excellence evaluation.

2. Teaching Evaluation

Teaching evaluations are usually conducted for two reasons; to improve practice and to assist in the faculty promotion process. However, the most important benefit lecturers can gain is feedback to help them refine their courses and teaching practices to provide students with better learning experiences [13]. Teaching evaluation is important in the refining of the teaching excellence of any lecturer. Another important reason for teaching evaluation is the improvement of class pass rate and thus the faculty's throughput. Although evaluation cannot be directly linked to throughput, it can assist lecturers to improve their teaching styles and upgrade course materials, which invariably impact on students' performances.

Taking this into consideration, lecturers should conduct three types of evaluations, namely students' evaluations, peer evaluations by colleagues, and evaluations by international guests and experts.

2.1. Students' Evaluation

Students benefit equally from proper classes and from research, and lecturers should be well prepared on both fronts. Descriptions and evaluations of both types of teaching evaluations are described below.

2.1.1. Students Observe Lecturers (SOL)

Students Observe lecturers (SOL) is an application designed and developed by the author in order to give students the ability to observe the lecture flow and send comments live to the lecturer in order to adjust their lecture's speed and flow. The designed App is installed on both the lecturer's and the students' computers. Students' numbers and their computers' IP addresses are considered in the App in order to secure the communication between students and lecturers within the class session. Many students are shy by nature and do not have the courage to ask a lecturer in the middle of the lecture. Also, some students are afraid to ask questions or to stop a lecturer to ask questions. The author, from his teaching experience, has realized that some students lose focus in the middle of the lecture due to disruption or the speed that the lecturer follows. To take control and give students the chance to slow down or catch up with the lecture, an application was designed and installed on the computers of each student in which they can click on different option to evaluate the flow of the lecture. Results will appear instantaneously on the screen of the lecturer's computer, who should check it from time to time to get an idea about the flow and the response from the students. Figures 3 and 4 show samples of the proposed SOL App.

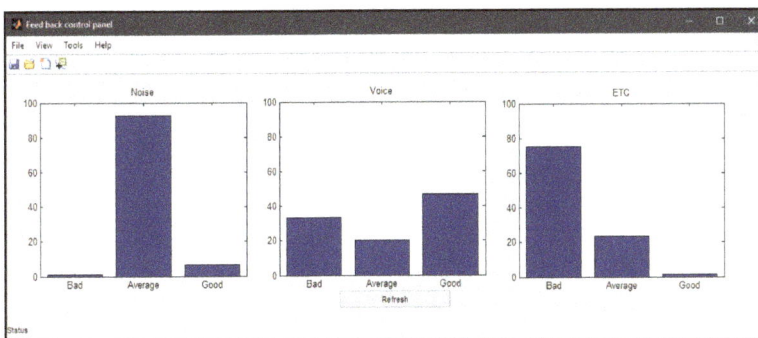

Figure 3. Students observe lecturers (SOL) feedback to the lecturer.

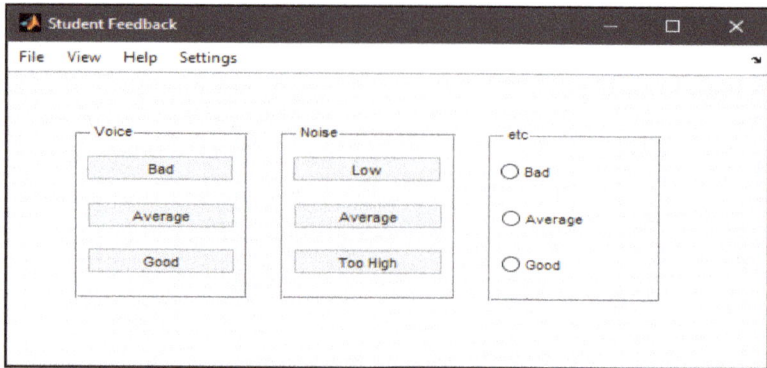

Figure 4. SOL students' comments.

2.1.2. Course Evaluation

Teaching evaluations from students have two major goals. The first is to assess the performance and teaching quality of lecturers and to provide them with insight on what they are doing well and how they need to improve. The second is to develop students' responsibilities towards their faculty through taking part in the evaluation process in order to improve teaching quality in the faculty.

Considering the above, lecturers should conduct surveys in which students participate anonymously to evaluate their lecturers' courses. In this paper, the author, who is lecturing a third year course on signal processing (SIG3B01), has conducted a survey which was based on a questionnaire which elicited students' overall opinions of the course, and specifically of the related course material such as slides, tutorial, practicals, and prescribed textbooks. Other questions were related to their opinions about the type of assessments and practicals on offer, and to the way in which their marks are calculated. Figure 5 shows the students' evaluation of the courses that he teaches.

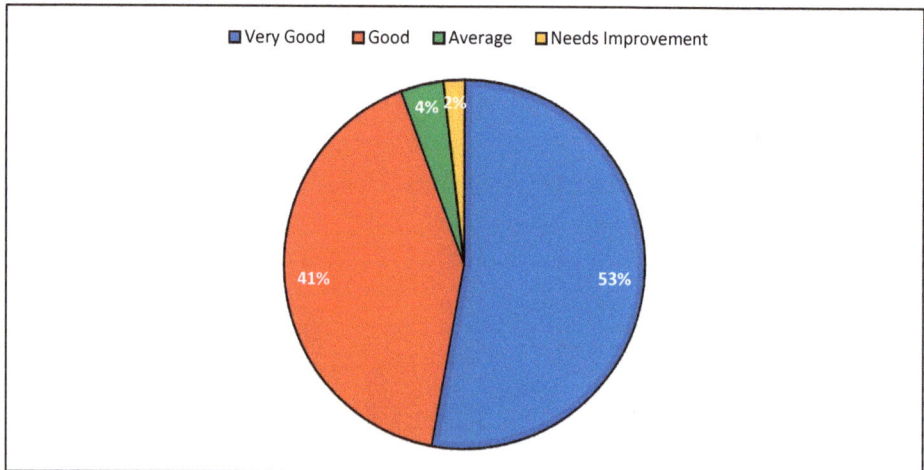

Figure 5. Students' evaluation of the SIG3B01 module.

2.1.3. Research Evaluation

In the author's experience, effective teaching is predicated on guidance and research; students need to be trained in both research methods and problem solving to be considered properly educated. The following are examples from different categories of students whom the author has had the pleasure of supervising in undergraduate, masters, and doctoral degrees. Figures 6 and 7 are samples of their personal evaluations to his research supervision.

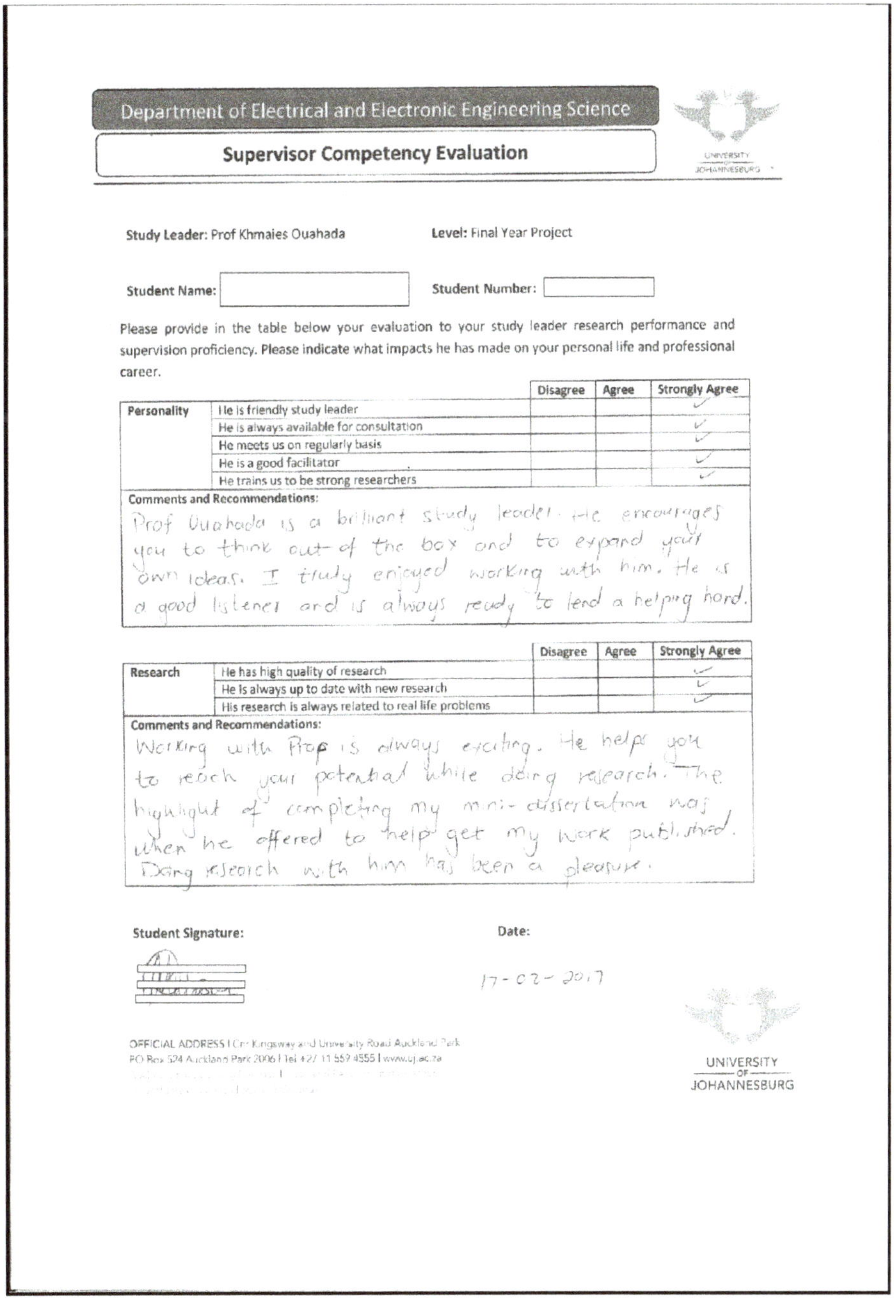

Department of Electrical and Electronic Engineering Science

Supervisor Competency Evaluation

UNIVERSITY OF JOHANNESBURG

Study Leader: Prof Khmaies Ouahada **Level:** Final Year Project

Student Name: **Student Number:**

Please provide in the table below your evaluation to your study leader research performance and supervision proficiency. Please indicate what impacts he has made on your personal life and professional career.

		Disagree	Agree	Strongly Agree
Personality	He is friendly study leader			✓
	He is always available for consultation			✓
	He meets us on regularly basis			✓
	He is a good facilitator			✓
	He trains us to be strong researchers			✓

Comments and Recommendations:
Prof Ouahada is a brilliant study leader. He encourages you to think out of the box and to expand your own ideas. I truly enjoyed working with him. He is a good listener and is always ready to lend a helping hand.

		Disagree	Agree	Strongly Agree
Research	He has high quality of research			✓
	He is always up to date with new research			✓
	His research is always related to real life problems			✓

Comments and Recommendations:
Working with Prof is always exciting. He helps you to reach your potential while doing research. The highlight of completing my mini-dissertation was when he offered to help get my work published. Doing research with him has been a pleasure.

Student Signature: **Date:** 17-02-2017

OFFICIAL ADDRESS | Cnr Kingsway and University Road Auckland Park
PO Box 524 Auckland Park 2006 | Tel +27 11 559 4555 | www.uj.ac.za

UNIVERSITY OF JOHANNESBURG

Figure 6. Research evaluation by undergraduate final year project student.

Department of Electrical and Electronic Engineering Science

Supervisor Competency Evaluation

UNIVERSITY OF JOHANNESBURG

Study Leader: Prof Khmaies Ouahada **Level:** Masters Degree

Student Name: **Student Number:**

Please provide in the table below your evaluation to your study leader research performance and supervision proficiency. Please indicate what impacts he has made on your personal life and professional career.

		Disagree	Agree	Strongly Agree
Personality	He is friendly study leader			✓
	He is always available for consultation			✓
	He meets us on regularly basis			✓
	He is a good facilitator			✓
	He trains us to be strong researchers			✓

Comments and Recommendations:

Prof. Khmaies is an excellent study leader. He helped me think out of the box and grow in my research capabilities. He provided invaluable input and feedback that helped me be more excellent in how I executed assigned tasks. He adequately prepared me for professional career.

		Disagree	Agree	Strongly Agree
Research	He has high quality of research			✓
	He is always up to date with new research			✓
	His research is always related to real life problems		✓	

Comments and Recommendations:

Prof. Khmaies is a well seasoned academic professional. He was full of new research ideas and he helped me grow in my research capabilities. I highly recommend him as a supervisor.

Student Signature: **Date:** 21/02/2017

OFFICIAL ADDRESS I Cnr Kingsway and University Road Auckland Park
PO Box 524 Auckland Park 2006 I Tel +27 11 559 4555 I www.uj.ac.za
Auckland Park Bunting Campus I Auckland Park Kingsway Campus
Doornfontein Campus I Soweto Campus

UNIVERSITY OF JOHANNESBURG

Figure 7. Research evaluation by Master's student.

2.2. Peer-Teaching Evaluation

A lecturer-to-lecturer evaluation is a means of obtaining accurate information about a colleague based on the fact that an evaluation from someone with experience in the same field and who knows the lecturer's work, ethic, and behavior would result in an ultimately more useful and accurate evaluation. This also has the potential to develop lecturers' working practices and help them understand the points of view of their colleagues [14–16].

Other benefits of lecturer-to-lecturer evaluation is the building of good working relationships between colleagues which will create a best practice environment throughout the university.

A few colleagues from different departments, universities, and different countries were invited to evaluate the author's teaching, as shown in Table 1.

Table 1. Peer-teaching evaluators.

Course	South African Universities		International University
	Author's University	Local University	
SIG3B01	A colleague from another department, Civil Engineering Science, University of Johannesburg, was invited to evaluate the author's teaching performance.	A colleague from another university, School of Electrical Engineering, University of the Witwatersrand, Johannesburg, was invited to evaluate the author's teaching	A colleague from an international university Duisburg-Essen University Germany, was invited to evaluate the author's teaching

2.2.1. Local-Teaching Evaluation

Considering the benefits of peer teaching evaluation mentioned above, the author asked colleagues from different schools in his university and other colleagues from other universities to evaluate his teaching performance and to provide feedback.

The author has chosen the heads of departments from schools of electrical engineering and civil engineering and asked them to attend his lectures to evaluate his teaching styles and his course material. He also asked the head of the Department of Electrical and Electronic Engineering Technology and the head of the Department of Civil Engineering Science from his university, the University of Johannesburg. A questionnaire evaluating the quality of his teaching was given to them. A similar questionnaire was given to another colleague from another university in South Africa at the School of Electrical Engineering at the University of Witwatersrand.

The reasons behind his choices were simple. Firstly, he needed feedback from a colleague in the same field of expertise and same school who is familiar with his curriculum and internal policies, as was the case with the Department of Electrical and Electronic Engineering Technology. Secondly, he also needed the opinion of someone who was from the same faculty but from a different school with different curricula, as was the case with the Department of Civil Engineering Science.

Thirdly, he needed an evaluation by a colleague from another university with different curricula and engineering programmes but from the same field of expertise, as was the case with the School of Electrical Engineering at the University of Witwatersrand. A sample of the questionnaire and evaluation form is presented in Figure 8.

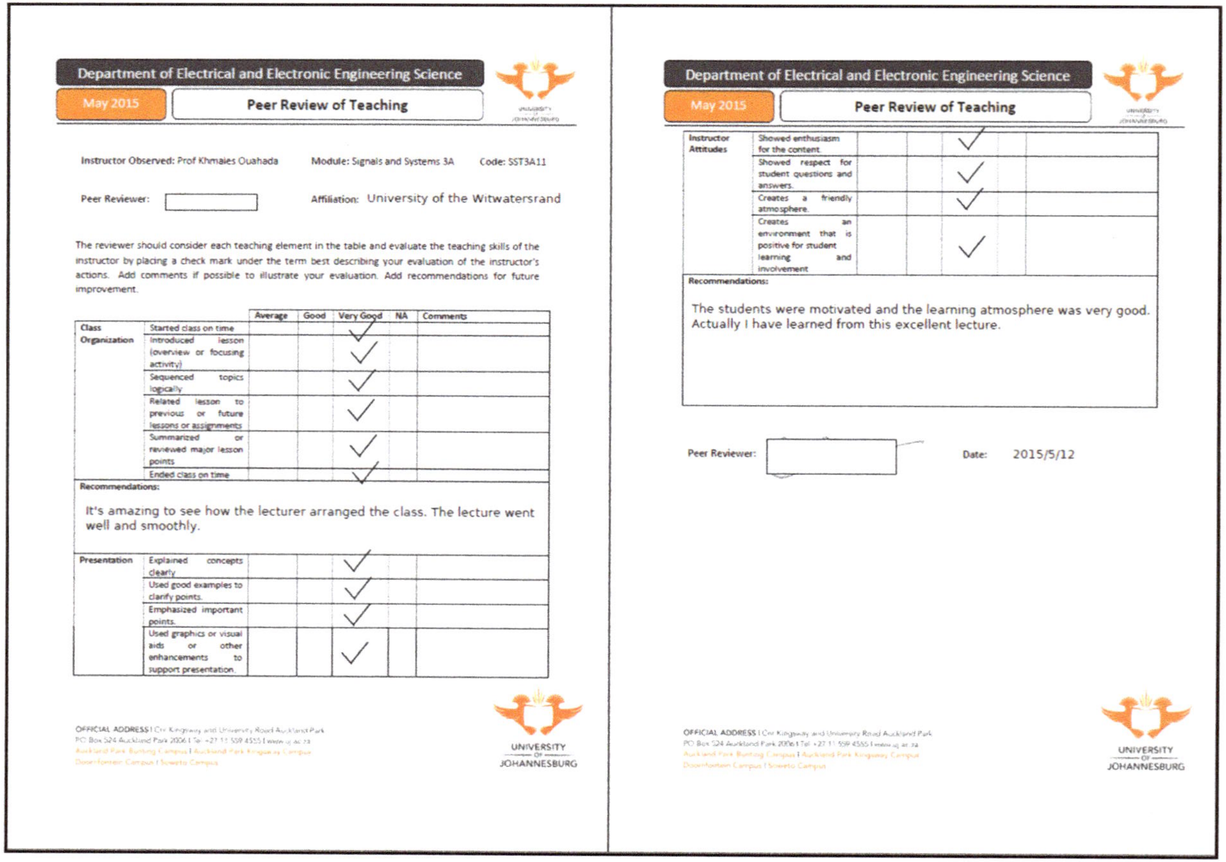

Department of Electrical and Electronic Engineering Science

May 2015 | Peer Review of Teaching

UNIVERSITY OF JOHANNESBURG

Instructor Observed: Prof Khmaies Ouahada   Module: Signals and Systems 3A   Code: SST3A11

Peer Reviewer: Affiliation: University of the Witwatersrand

The reviewer should consider each teaching element in the table and evaluate the teaching skills of the instructor by placing a check mark under the term best describing your evaluation of the instructor's actions. Add comments if possible to illustrate your evaluation. Add recommendations for future improvement.

		Average	Good	Very Good	NA	Comments
Class Organization	Started class on time			✓		
	Introduced lesson (overview or focusing activity)			✓		
	Sequenced topics logically			✓		
	Related lesson to previous or future lessons or assignments			✓		
	Summarized or reviewed major lesson points			✓		
	Ended class on time			✓		

Recommendations:

It's amazing to see how the lecturer arranged the class. The lecture went well and smoothly.

		Average	Good	Very Good	NA	Comments
Presentation	Explained concepts clearly			✓		
	Used good examples to clarify points.			✓		
	Emphasized important points.			✓		
	Used graphics or visual aids or other enhancements to support presentation.			✓		

OFFICIAL ADDRESS | Cnr Kingsway and University Road Auckland Park
PO Box 524 Auckland Park 2006 | Tel +27 11 559 4555 | www.uj.ac.za
Auckland Park Bunting Campus | Auckland Park Kingsway Campus
Doornfontein Campus | Soweto Campus

UNIVERSITY OF JOHANNESBURG

Department of Electrical and Electronic Engineering Science

May 2015 | Peer Review of Teaching

		Average	Good	Very Good	NA	Comments
Instructor Attitudes	Showed enthusiasm for the content.			✓		
	Showed respect for student questions and answers.			✓		
	Creates a friendly atmosphere.			✓		
	Creates an environment that is positive for student learning and involvement			✓		

Recommendations:

The students were motivated and the learning atmosphere was very good. Actually I have learned from this excellent lecture.

Peer Reviewer: Date: 2015/5/12

OFFICIAL ADDRESS | Cnr Kingsway and University Road Auckland Park
PO Box 524 Auckland Park 2006 | Tel +27 11 559 4555 | www.uj.ac.za
Auckland Park Bunting Campus | Auckland Park Kingsway Campus
Doornfontein Campus | Soweto Campus

UNIVERSITY OF JOHANNESBURG

Figure 8. Peer evaluation from the University of the Witwatersrand.

2.2.2. International Teaching Evaluation

In line with the earlier explanation, lecturers should try to get an international evaluation regarding their course materials and lecturing skills. In this context, the author took the opportunity to invite two international professors on two different occasions to attend his lectures to get their opinions about his teaching style, course material, and the teaching environment he provides to his students.

The author had the opportunity to invite a professor from Armenia. A sample of the questionnaire and evaluation is presented in Figure 9.

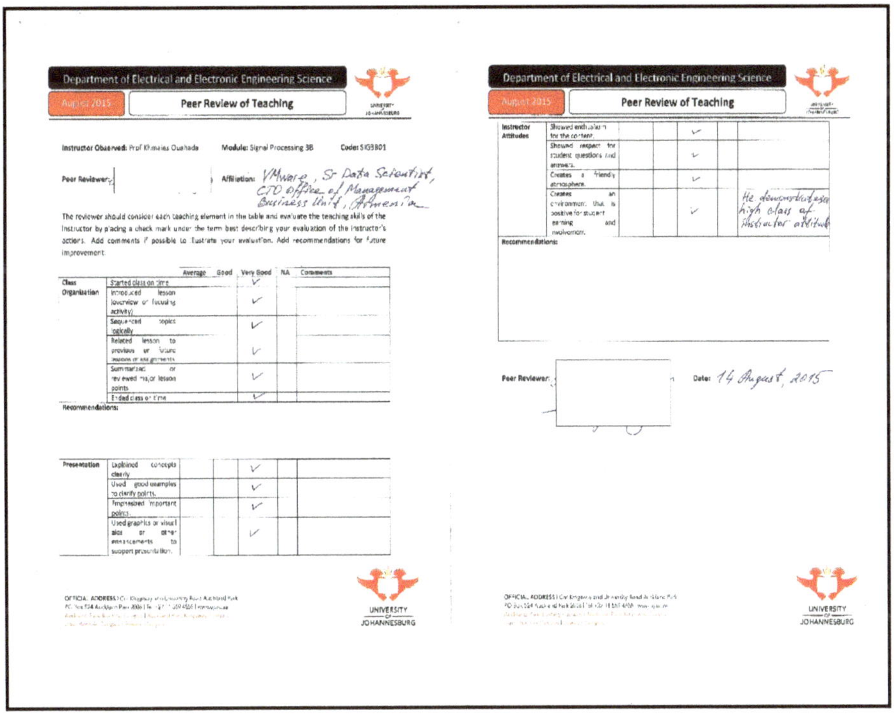

Figure 9. Peer evaluation from VMware, Armenia.

2.2.3. Self-Teaching Evaluation

As academics, lecturers should acquire benefits from overseas institutions either via research collaborations or by attending conferences. These are very important opportunities to gain exposure to different professional environments and to develop communication skills through interactions with academics from different parts of the world. Exploring opportunities to lecture at international universities and to get feedback from their students and staff is also very important in the professional career of an academic. The author had the opportunity to teach a course in 2014 as a visiting researcher at the University of Duisburg-Essen in Germany. The faculty of engineering approached him to design and lecture a Master's degree course on Information Theory. Feedback from the University of Duisburg-Essen about his teaching experience is presented in Figure 10.

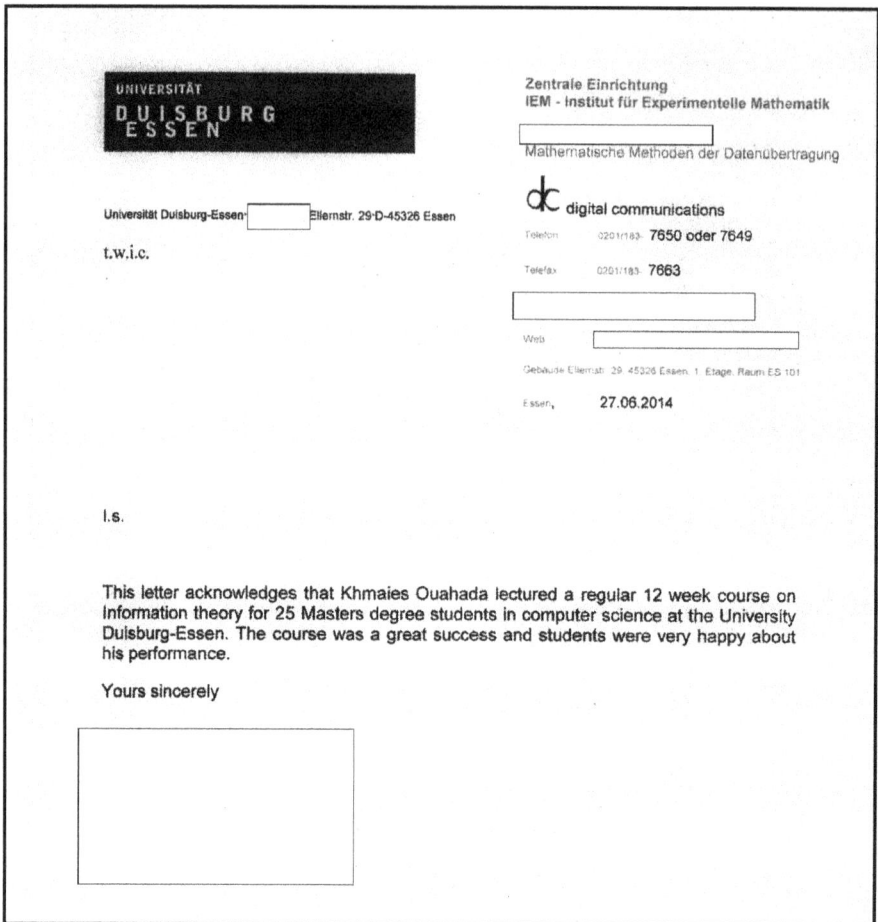

Figure 10. Evaluation letter from Duisburg-Essen University, Germany.

3. Course Material

This section discusses the university's moderation philosophy and presents evidence of its practice. Universities usually stipulate a formal moderation procedure. In this paper, the author has extended this policy by approaching students and local and international colleagues to evaluate his courses for the sake of enhancing his teaching expertise.

In his case, the author's faculty curriculum comprises two types of modules or courses, namely core/fundamental modules and exit-level modules. The Engineering Council South Africa (ECSA), a watchdog for engineering curriculum quality, call final degree courses exit-level outcomes (ELO). These courses are assessed and moderated twice in the exit level modules, both internally and externally by someone from another university. This entails proper moderation of all the exam papers, module content, and answer sheets. Non-exit-level modules can be moderated internally by colleagues as is the case with the author's own modules, which are third year modules. Files with all course material evidence, called ECSA files, are prepared by lecturers and moderators are expected to inspect these for evidence of course teaching.

3.1. Students' Moderation

3.1.1. Textbooks

The recommended textbook is a further benefit to students that provides an alternative solution to students who could not afford buying the prescribed textbook in the first place. Although the prescribed textbooks are available in the university's libraries, many students are unable to access them because of the limited copies in relation to the number of students registered for the course.

From experience, another way in which lecturers could help students, mostly those who are struggling financially and cannot afford very expensive engineering textbooks, is for them to design their own textbooks. In this regard, the author designed his own textbook and made it available to his students on their blackboard, a website for course management. This e-book was inspired by the recommended textbook and was intended to build a strong relationship between the slides of his lecture notes and the prescribed textbook.

The author makes a direct link between the content of his slides used in class and the corresponding content in the prescribed textbook. The result was very positive and students liked the idea and his e-book became more readable and accessible to students than other available textbooks.

The author believes lecturers should always obtain feedback by getting their students to evaluate lecturers' initiatives and ideas. In this regard, he conducted a survey among his students to gauge their responses to the prescribed textbook, the recommended textbook, and his own textbook. Figures 11–13 show the reaction of his students to these aspects.

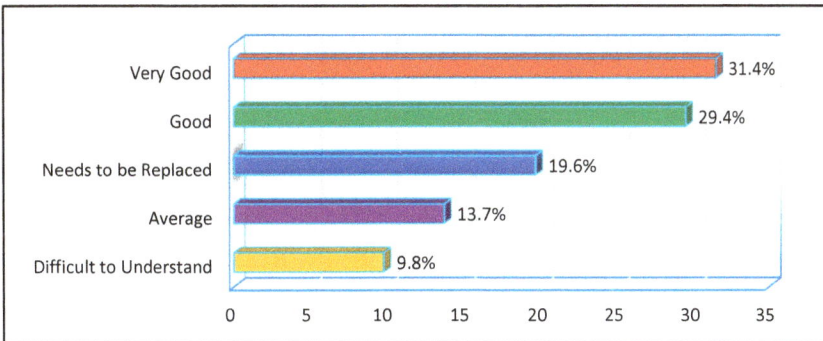

Figure 11. Students' Evaluation of the prescribed textbook.

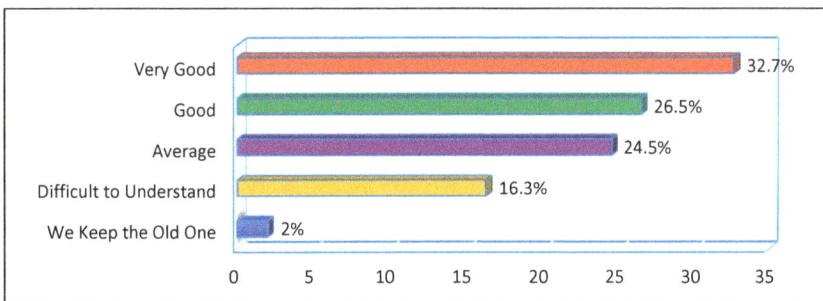

Figure 12. Students' evaluation of the recommended textbook.

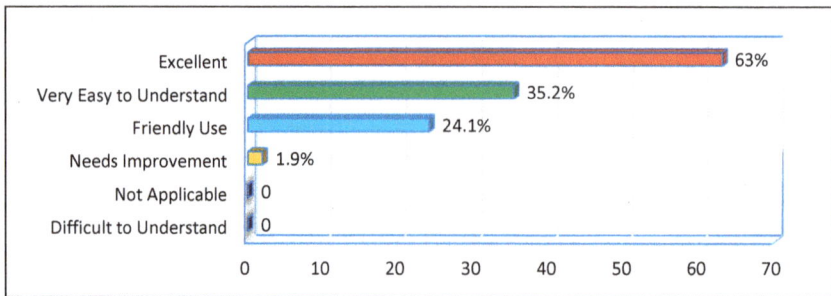

Figure 13. Students' evaluation of the author's proposed e-Book.

3.1.2. Lecture Notes

From the author's experience in teaching, the author always designs his notes to be structured in a manner that would promote a logical and elucidatory flow to help students who "get lost" in the lecture and gives them a chance to catch up. This is achieved by inserting short problems in between subsections, where students are asked to solve them. These questions will help students to understand the previous slides' contents and help others to catch up during the time reserved for such applications. At the end of each lecture, an example ought to be given that summarizes all sections in the slides and promotes clear understanding of the lecture.

Figure 14, results from a survey that was conducted among the author's students, shows their reactions, their opinions, and their ratings of the author's lecture notes and slides.

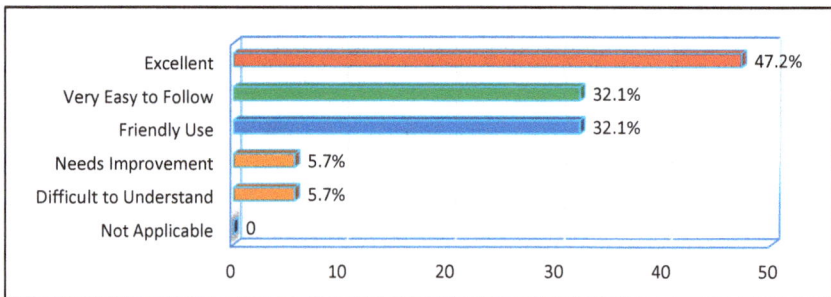

Figure 14. Students' evaluation of lecture notes.

3.1.3. Tutorials

From the author's lecturing experience, the best way to teach students is to create a competitive atmosphere among them and give prizes. Students love competing with each other, especially when rewards are on offer. In order to help students arrive at the correct answers, the lecturer should ask relevant questions. Brain-storming challenging questions makes students feel that they are in class not just to take notes and leave, but to take part in finding solutions and being proud of that achievement. Students should know that active participation in class and tutorials will ease their revision at home.

A survey was conducted among the author's students to gauge their opinions and rating of his tutorials. The results are shown in Figure 15.

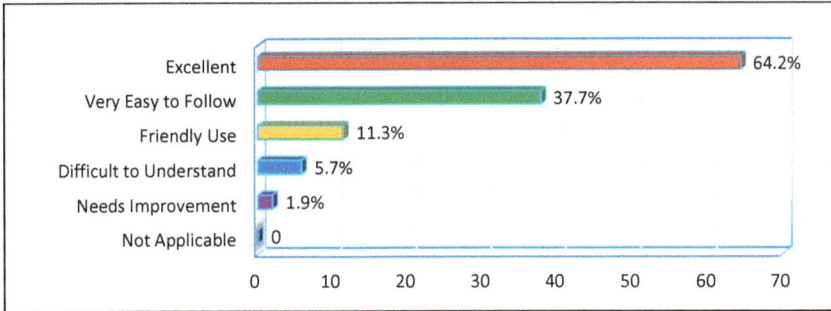

Figure 15. Student evaluation of tutorials.

3.1.4. Practicals

Lecturers should pay attention to the capacity of students to handle practicals [17–19].

It is very important to recognize that students come from different backgrounds and possess different skills as a result of their secondary educational experiences. Many students come from underprivileged places where it is hard and even impossible to get access to computers and thus programming skills differ from one student to the other. The same is applicable to hardware practicals.

Based on the above, a practical that caters for different types of students will be fair for all of them. In this regard, a survey was conducted among the author's students to gauge their opinions on the quality of his practicals and their preference of the type of practicals they prefer—hardware, software, or a combination of both. Interestingly, although programming was the least preferable choice for practicals, students accepted it with hardware implementation because they know that engineers ought to improve their programming skills to be ready for industry.

Figures 16 and 17 respectively show, firstly, the ratings of his students regarding the quality of his practicals and, secondly, their preferences regarding the three different types of practicals and their best choices.

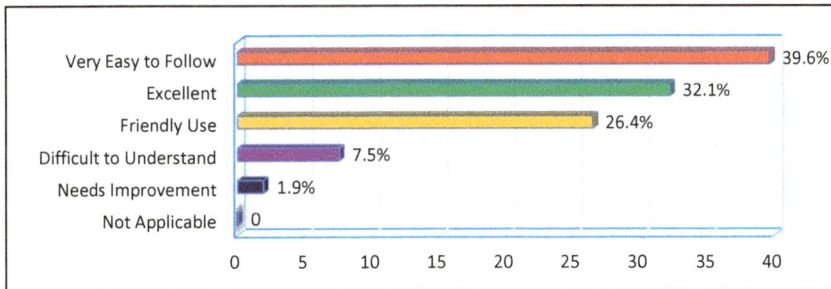

Figure 16. Students' evaluation of the practicals.

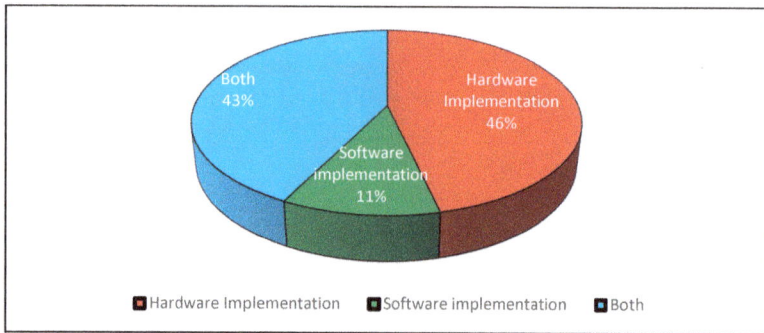

Figure 17. Students' preferences for different types of practicals.

3.1.5. Class Tests

Reduced student workloads in a stress-free assessment atmosphere helps to improve throughput rates. This can be achieved by designing assessment schemes which allow both students and lecturers more time and flexibility to prepare for courses.

The department of Electrical and Electronic Engineering Science has changed its assessment strategy from the traditional summative assessment model consisting of semester tests and exams to a more fine-tuned outcomes-based continuous assessment model [20]. Each module is divided into a set of outcomes which encompass key knowledge areas and can be regarded as a chapter with a common theme. This system is considered to offer optimum efficiency for knowledge acquisition and serves to demonstrate our students' capabilities to pass all knowledge areas in each module. During the course of the semester, students are given three small formative assessments for each module outcome. To pass a module outcome, a student has to achieve a 50% mark in two of the assessments or a 70% mark in one of the assessments.

The philosophy is that a student can fail one opportunity and use the experience gained to pass subsequent assessments. The 70% threshold was instituted to allow students who have mastered given outcomes the opportunity to demonstrate their knowledge once and then be able to focus on the remaining work.

The author has been using outcome-based assessments since 2011. In 2015, he had the opportunity to lecture three courses to third-year students, namely Signal and Systems (SST3A11) in the first semester, and Digital Signal Processing (SIG3B01) and Telecommunications (TEL3B01) in the second semester. After almost five years of using outcome-based assessments, the author decided in 2015 to evaluate this assessment scheme and to develop an assessment scheme using different assessment styles. This was done to avoid the heavy load caused by the outcome-based assessment and the types of questions given to students.

In his first semester course, Signal and Systems, the author applied the departmental assessment module, treating the practicals as outcomes on their own. The scheme for calculating student marks is depicted in Table 2.

In the second semester of the SIG3B01 module, the author applied a different assessment scheme from the one used with the SST3A11 module: he retained the three assessment opportunities to meet the ECSA requirements but treated the practicals as one of the assessments. The reason for dropping the number of assessments was because they create a heavy load on students, affect their results, and therefore the throughput rate. Another modification was to give different varieties of assessment that did not only focus on problem solving and derivation. He introduced a multiple-choice type of assessment to cater for students who are not comfortable with problem solving as the only type of assessment. Since different types of questions require different time allocations, he adjusted the percentage of each assessment mark to the final mark. The final assessment weights are shown in Table 3.

Table 2. Scheme used for assessment of the SST3A11 module.

Assessments	Kind of Assessment	Assessment Details	Assessment Weight	Outcome Weight
		Outcome A		25%
Assessment 1	Writing assessment	Problem Solving and Derivation	70% Exemption 0.7 Max 1 + 0.3 Max 2	
Assessment 2	Practical	Problem Solving and Derivation		
Assessment 3	Writing assessment	Problem Solving and Derivation		
		Outcome B		25%
Assessment 1	Writing assessment	Problem Solving and Derivation	70% Exemption 0.7 Max 1 + 0.3 Max 2	
Assessment 2	Practical	Problem Solving and Derivation		
Assessment 3	Writing assessment	Problem Solving and Derivation		
		Outcome C		25%
Assessment 1	Writing assessment	Problem Solving and Derivation	70% Exemption 0.7 Max 1 + 0.3 Max 2	
Assessment 2	Practical	Problem Solving and Derivation		
Assessment 3	Writing assessment	Problem Solving and Derivation		
		Outcome D		25%
		Practicals: Reports and Matlab programming		
		Final Mark		
	Average (Outcome A + Outcome B + Outcome C + Outcome D)			100%

Table 3. Structure used for assessments of the SIG3B01 module.

Assessments	Kind of Assessment	Assessment Details	Assessment Weight	Outcome Weight
		Outcome A		33%
Assessment 1	Writing assessment	Multiple-Choice + Theory	30%	
Assessment 2	Practical	Report + Demonstration	30%	
Assessment 3	Writing assessment	Problem Solving and Derivation	40%	
		Outcome B		33%
Assessment 1	Writing assessment	Multiple-Choice + Theory	30%	
Assessment 2	Practical	Report + Demonstration	30%	
Assessment 3	Writing assessment	Problem Solving and Derivation	40%	
		Outcome C		33%
Assessment 1	Writing assessment	Multiple-Choice + Theory	30%	
Assessment 2	Practical	Report + Demonstration	30%	
Assessment 3	Writing assessment	Problem Solving and Derivation	40%	
		Final Mark		
	Average (Outcome A + Outcome B + Outcome C)			100%

In the case of the TEL3B01 module, the author kept the same assessment style as for SIG3B01 but moved the practicals on their own—not as in the case of SST3A11—into a small project which contributed a certain percentage to the final mark of the module. The idea behind this was to give students a chance to do a separate project for submission at the end of the semester while taking advantage of the practical allocated times to do revision or homework. This new model, to be consistent with the ECSA's assessment requirement, needed a third assessment. The author thus introduced a quiz which carried a lower percentage to accommodate the rest of the assessment types. The final assessment weights are shown in Table 4.

Table 4. Scheme used for the TEL3B01 module assessment.

Assessments	Kind of Assessment	Assessment Details	Assessment Weight	Outcome Weight
Outcome A				35%
Assessment 1	Test	Quiz	20%	
Assessment 2	Test	Multiple-Choice + Theory	30%	
Assessment 3	Test	Problem Solving and Derivation	50%	
Outcome B				35%
Assessment 1	Test	Quiz	20%	
Assessment 2	Test	Multiple-Choice + Theory	30%	
Assessment 3	Test	Problem Solving and Derivation	50%	
Practical				30%
Practical	Project	Report	30%	
		Hardware implementation	70%	
Final Mark				
0.35 × Outcome A + 0.35 × Outcome B + 0.3 × Practical				100%

The author conducted a survey among his students to gauge their preferences and to assess how comfortable they were with each of the assessment types. He preferred not to rely on the results only but wanted them to express their views on this matter to assist in improving the proposed assessment schemes.

Figures 18–20 provide information on student choices and the type of scheme that helped to improve their marks. Figure 18 illustrates that students prefer the assessments tool used for SIG3B01 as the scheme that best suits an outcomes-based approach. From Figure 19, it is clear that the assessment tool used with SIG3B01 is the one they feel comfortable with. Figure 20 shows that multiple-choice assessment is the best tool to help students improve their marks.

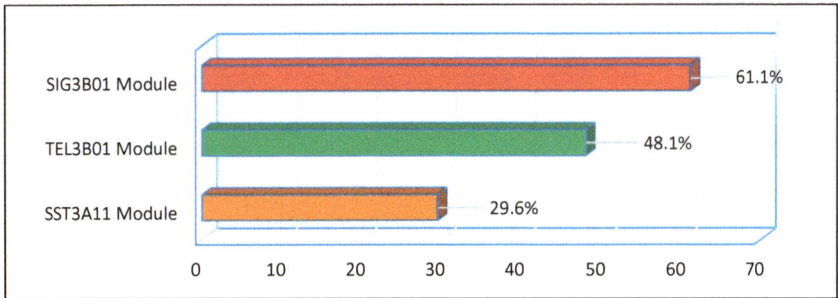

Figure 18. Student evaluation of module assessments schemes.

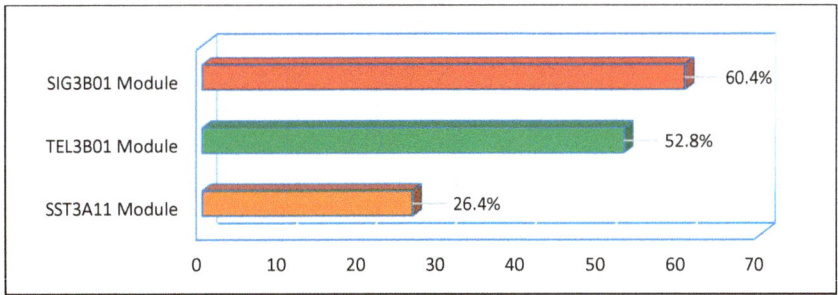

Figure 19. Student preferences for the assessment schemes.

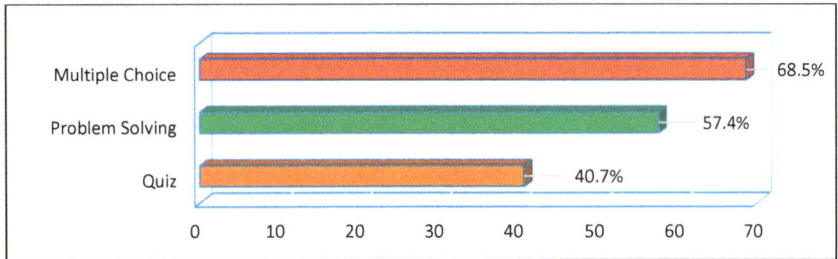

Figure 20. Student preferences for different types of assessment.

3.2. Professional Moderation

This section offers an explanation of the author's moderation philosophy in the context of continuous evaluation of his course content by local as well as national and international colleagues, at his request. Table 5 gives an overview of his approach, and the statistics that are offered as evidence of his moderation.

Table 5. Peer-module moderation.

Courses	Course Material	Course Moderation		International University
		South African University		
		Author's University	Local University	
SIG3B01	• Lecture Notes • Tutorial • Practicals • Textbook	Colleagues from other departments were invited to evaluate the lecturer's course material. The line manager as the head of the department (HOD) took part in this process.	Colleagues from other universities were invited to evaluate the lecturer's course material.	Colleagues from other countries and international universities were invited to evaluate the lecturer's course material.

The author has approached colleagues and academics from different departments, universities, and countries to moderate his course material, as shown in Table 5. He tried to get a broader opinion from academics from different backgrounds. He has asked colleagues from different departments of his faculty of engineering. He has also asked the opinion from a line management perspective from his head of department. He has approached colleagues from the University of the Witwatersrand, which has very strong ties with industry, and the University of Pretoria, which has longer history than his university. At the international level, he has approached academics from North Africa (Tunisia) and from Europe and Asia (Italy and Oman).

The author has personally prepared a special moderation form, which was sent to the moderators with his course containing his study guide, practical guide, lecture notes, tutorials, and his textbook. Figures 21–23 are examples of moderation reports by academics from local and international universities. These moderation forms were designed by himself with questions to moderate the course material as explained earlier.

This kind of moderation is informed by his belief that lecturers should challenge themselves to induce innovative thinking. It also builds self-confidence to know that your course material has been subjected to expert evaluation and acknowledged by colleagues whom you respect both locally and internationally.

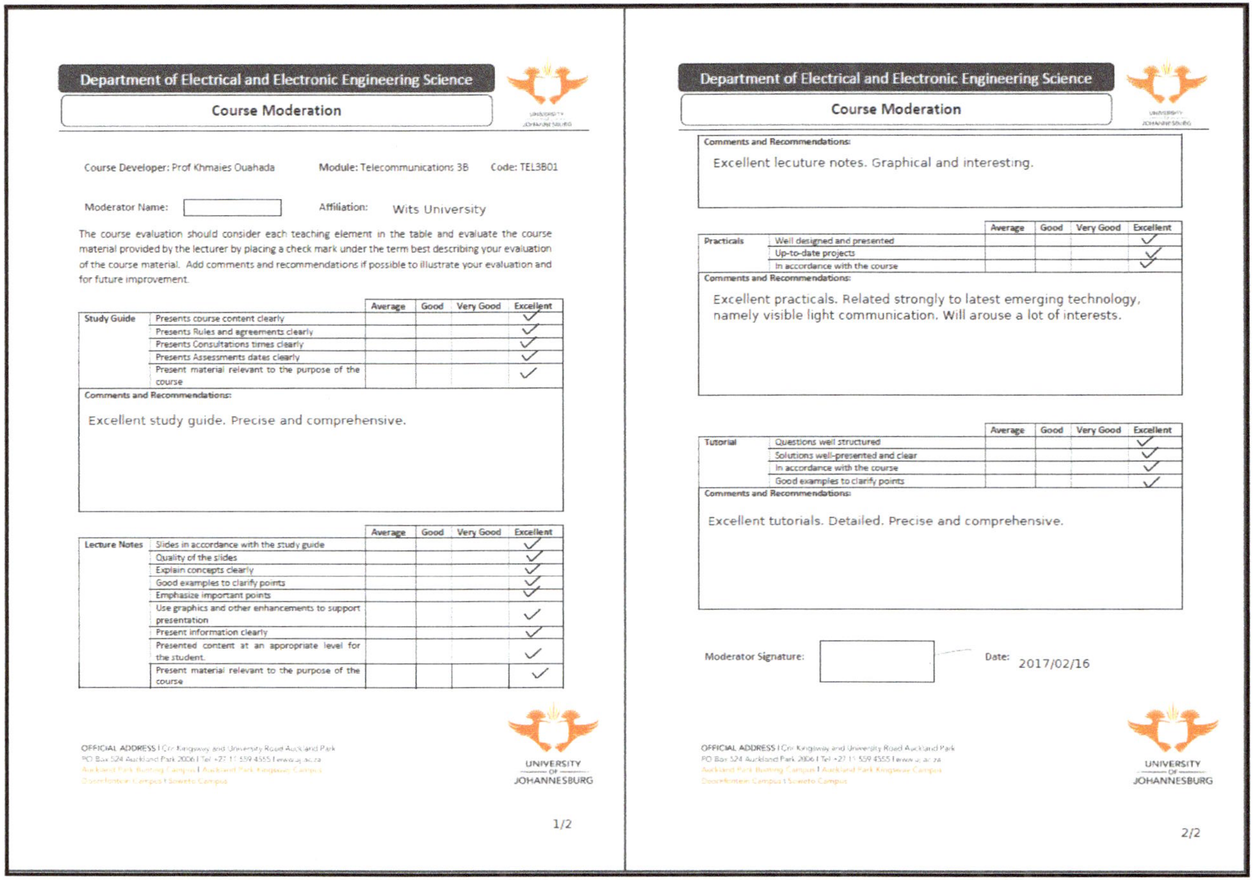

Department of Electrical and Electronic Engineering Science

Course Moderation

Course Developer: Prof Khmaies Ouahada Module: Telecommunications 3B Code: TEL3B01

Moderator Name: Affiliation: Wits University

The course evaluation should consider each teaching element in the table and evaluate the course material provided by the lecturer by placing a check mark under the term best describing your evaluation of the course material. Add comments and recommendations if possible to illustrate your evaluation and for future improvement.

		Average	Good	Very Good	Excellent
Study Guide	Presents course content clearly				✓
	Presents Rules and agreements clearly				✓
	Presents Consultations times clearly				✓
	Presents Assessments dates clearly				✓
	Present material relevant to the purpose of the course				✓

Comments and Recommendations:

Excellent study guide. Precise and comprehensive.

		Average	Good	Very Good	Excellent
Lecture Notes	Slides in accordance with the study guide				✓
	Quality of the slides				✓
	Explain concepts clearly				✓
	Good examples to clarify points				✓
	Emphasize important points				✓
	Use graphics and other enhancements to support presentation				✓
	Present information clearly				✓
	Presented content at an appropriate level for the student.				✓
	Present material relevant to the purpose of the course				✓

OFFICIAL ADDRESS | Cnr Kingsway and University Road Auckland Park

UNIVERSITY OF JOHANNESBURG

1/2

Department of Electrical and Electronic Engineering Science

Course Moderation

Comments and Recommendations:

Excellent lecuture notes. Graphical and interesting.

		Average	Good	Very Good	Excellent
Practicals	Well designed and presented				✓
	Up-to-date projects				✓
	In accordance with the course				✓

Comments and Recommendations:

Excellent practicals. Related strongly to latest emerging technology, namely visible light communication. Will arouse a lot of interests.

		Average	Good	Very Good	Excellent
Tutorial	Questions well structured				✓
	Solutions well-presented and clear				✓
	In accordance with the course				✓
	Good examples to clarify points				✓

Comments and Recommendations:

Excellent tutorials. Detailed. Precise and comprehensive.

Moderator Signature: Date: 2017/02/16

OFFICIAL ADDRESS | Cnr Kingsway and University Road Auckland Park

UNIVERSITY OF JOHANNESBURG

2/2

Figure 21. TEL3B01 module moderation report by a colleague from the School of electrical engineering at Wits University.

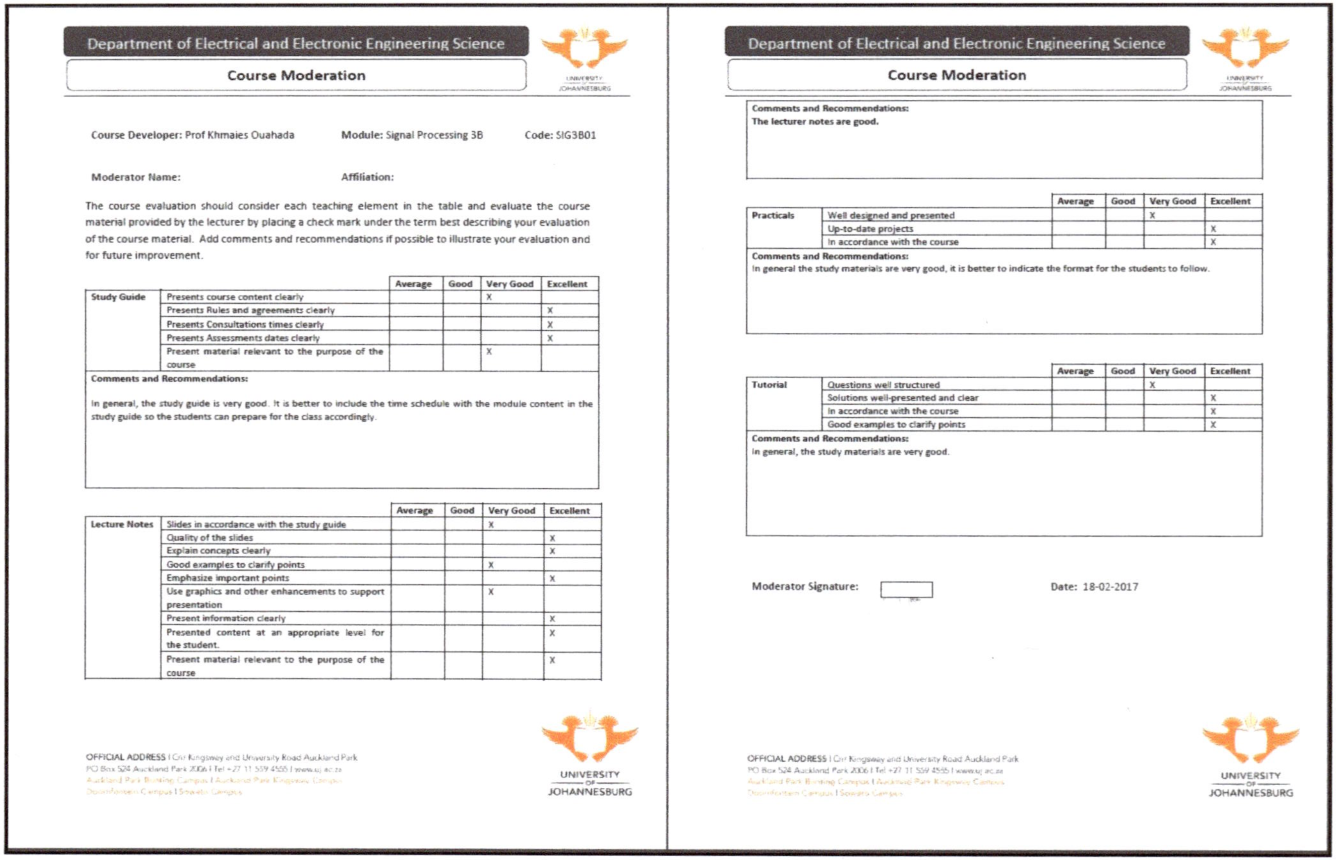

Department of Electrical and Electronic Engineering Science

Course Moderation

Course Developer: Prof Khmaies Ouahada   Module: Signal Processing 3B   Code: SIG3B01

Moderator Name: Affiliation:

The course evaluation should consider each teaching element in the table and evaluate the course material provided by the lecturer by placing a check mark under the term best describing your evaluation of the course material. Add comments and recommendations if possible to illustrate your evaluation and for future improvement.

		Average	Good	Very Good	Excellent
Study Guide	Presents course content clearly			X	
	Presents Rules and agreements clearly				X
	Presents Consultations times clearly				X
	Presents Assessments dates clearly				X
	Present material relevant to the purpose of the course			X	

Comments and Recommendations:

In general, the study guide is very good. It is better to include the time schedule with the module content in the study guide so the students can prepare for the class accordingly.

		Average	Good	Very Good	Excellent
Lecture Notes	Slides in accordance with the study guide			X	
	Quality of the slides				X
	Explain concepts clearly				X
	Good examples to clarify points			X	
	Emphasize important points				X
	Use graphics and other enhancements to support presentation			X	
	Present information clearly				X
	Presented content at an appropriate level for the student.				X
	Present material relevant to the purpose of the course				X

OFFICIAL ADDRESS | Cnr Kingsway and University Road Auckland Park
PO Box 524 Auckland Park 2006 | Tel +27 11 559 4555 | www.uj.ac.za
Auckland Park Bunting Campus | Auckland Park Kingsway Campus
Doornfontein Campus | Soweto Campus

UNIVERSITY OF JOHANNESBURG

Department of Electrical and Electronic Engineering Science

Course Moderation

Comments and Recommendations:
The lecturer notes are good.

		Average	Good	Very Good	Excellent
Practicals	Well designed and presented			X	
	Up-to-date projects				X
	In accordance with the course				X

Comments and Recommendations:
In general the study materials are very good, it is better to indicate the format for the students to follow.

		Average	Good	Very Good	Excellent
Tutorial	Questions well structured			X	
	Solutions well-presented and clear				X
	In accordance with the course				X
	Good examples to clarify points				X

Comments and Recommendations:
In general, the study materials are very good.

Moderator Signature: Date: 18-02-2017

OFFICIAL ADDRESS | Cnr Kingsway and University Road Auckland Park
PO Box 524 Auckland Park 2006 | Tel +27 11 559 4555 | www.uj.ac.za
Auckland Park Bunting Campus | Auckland Park Kingsway Campus
Doornfontein Campus | Soweto Campus

UNIVERSITY OF JOHANNESBURG

Figure 22. SIG3B module moderation report by the HOD.

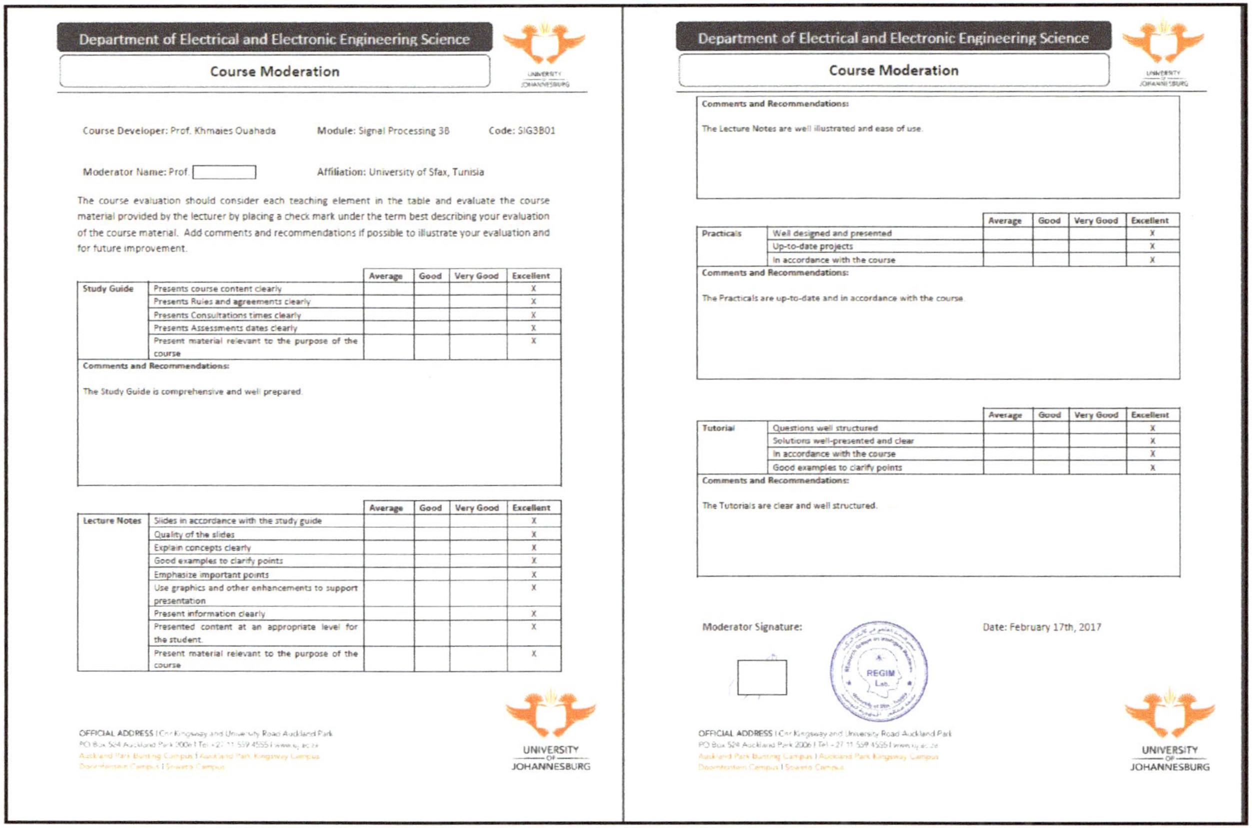

Department of Electrical and Electronic Engineering Science

Course Moderation

Course Developer: Prof. Khmaies Ouahada   Module: Signal Processing 3B   Code: SIG3B01

Moderator Name: Prof.   Affiliation: University of Sfax, Tunisia

The course evaluation should consider each teaching element in the table and evaluate the course material provided by the lecturer by placing a check mark under the term best describing your evaluation of the course material. Add comments and recommendations if possible to illustrate your evaluation and for future improvement.

		Average	Good	Very Good	Excellent
Study Guide	Presents course content clearly				X
	Presents Rules and agreements clearly				X
	Presents Consultations times clearly				X
	Presents Assessments dates clearly				X
	Present material relevant to the purpose of the course				X

Comments and Recommendations:

The Study Guide is comprehensive and well prepared.

		Average	Good	Very Good	Excellent
Lecture Notes	Slides in accordance with the study guide				X
	Quality of the slides				X
	Explain concepts clearly				X
	Good examples to clarify points				X
	Emphasize important points				X
	Use graphics and other enhancements to support presentation				X
	Present information clearly				X
	Presented content at an appropriate level for the student.				X
	Present material relevant to the purpose of the course				X

OFFICIAL ADDRESS | Cnr Kingsway and University Road Auckland Park
PO Box 524 Auckland Park 2006 | Tel +27 11 559 4555 | www.uj.ac.za
Auckland Park Bunting Campus | Auckland Park Kingsway Campus
Doornfontein Campus | Soweto Campus

UNIVERSITY OF JOHANNESBURG

Department of Electrical and Electronic Engineering Science

Course Moderation

Comments and Recommendations:

The Lecture Notes are well illustrated and ease of use.

		Average	Good	Very Good	Excellent
Practicals	Well designed and presented				X
	Up-to-date projects				X
	In accordance with the course				X

Comments and Recommendations:

The Practicals are up-to-date and in accordance with the course

		Average	Good	Very Good	Excellent
Tutorial	Questions well structured				X
	Solutions well-presented and clear				X
	In accordance with the course				X
	Good examples to clarify points				X

Comments and Recommendations:

The Tutorials are clear and well structured.

Moderator Signature: Date: February 17th, 2017

REGIM Lab.

OFFICIAL ADDRESS | Cnr Kingsway and University Road Auckland Park
PO Box 524 Auckland Park 2006 | Tel +27 11 559 4555 | www.uj.ac.za
Auckland Park Bunting Campus | Auckland Park Kingsway Campus
Doornfontein Campus | Soweto Campus

UNIVERSITY OF JOHANNESBURG

Figure 23. SIG3B module moderation report by a colleague from ENIS, Sfax, Sfax University, Tunisia.

4. Discussion

In this section, we present the benefits of course moderation and teaching evaluation and their effect on the students' performance and class pass rate. The feedback obtained through students and professionals is the key to improvement of teaching quality and class pass rate.

4.1. Course Evaluation

Students teaching evaluation has two major goals. The first is to evaluate the performance and teaching quality of their lecturers and provide them with insight on what they are doing well and how they need to improve. The second goal is to build within students' responsibility towards their university by taking part in the drive to improve teaching quality for the future enrolled students.

Taking into consideration the feedback from different surveys presented in the previous sections, the author conducted a survey among students to evaluate the overall of his three courses he lectured, Signal and systems (SST3A11), digital signals (SIG3B01) and analog telecommunications (TEL3B01).

Figures 24 and 25 are respectively students' evaluation for analog and digital signals and systems. The content of both courses is mainly on transforms and filters. It is clear from the figures that students' opinions/ratings are close for both courses, with a rate of higher than 90%. With regards to TEL3B01 module which deals with analog modulations, the extensive theory makes the course unattractive, different from Signals and Systems courses. Figure 26 shows students rating of more than 80%. These results show how important is for lecturers take into consideration the feedback from students and colleagues in order to improve his course.

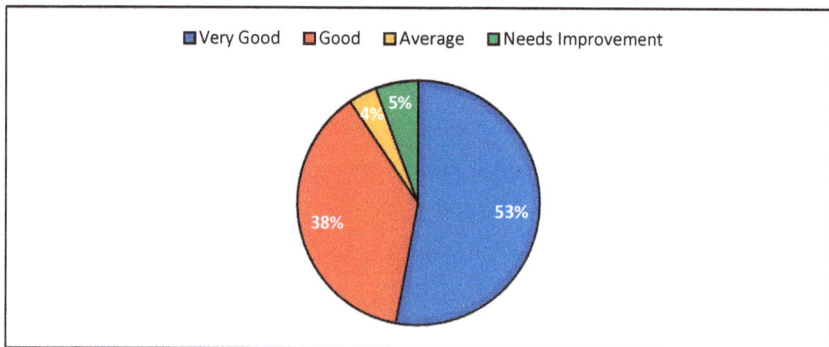

Figure 24. Student evaluation of the SST3A11 module.

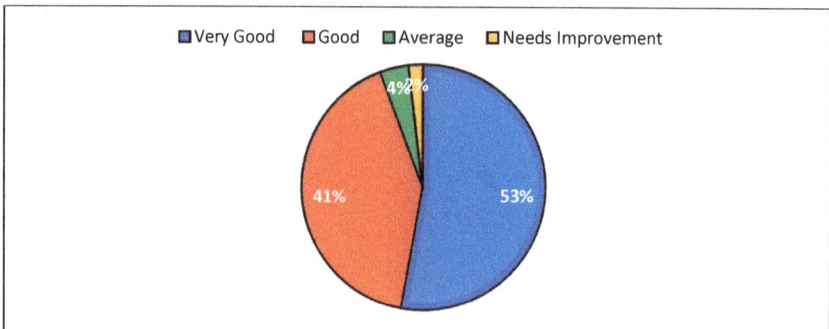

Figure 25. Student evaluation of the SIG3B01 module.

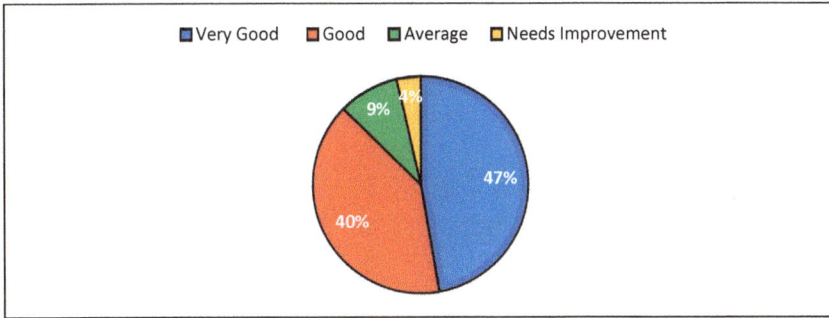

Figure 26. Students evaluation of the TEL3B01 module.

4.2. Pass Rate

Asking students and colleagues to evaluate your course helps improve the course because preparing a high-quality and friendly educational environment for students is one of the most important factors in improving class pass rate. Students' opinions and interaction with professionals from local or international universities help make a course more fruitful. Providing more consultation time and revision sessions to help students catch up with the course before examinations and helping underprivileged students with free hard copies of lecture notes play significant roles in making the course accessible to students.

The comparative results between the author's modules SST3A11, SIG3B01, and TEL3B01 lectured in 2015, and a summary of the benefit of the author's teaching philosophy in improving the class pass rates are presented in Figure 27.

Figure 27. Class pass rates of different modules.

Comparative results between the author's modules in SIG3B01 and TEL3B01 lectured since 2010 summarizing the benefit of improving the author's teaching skills and updating his teaching philosophy in improving the class pass rates are presented in Figure 28.

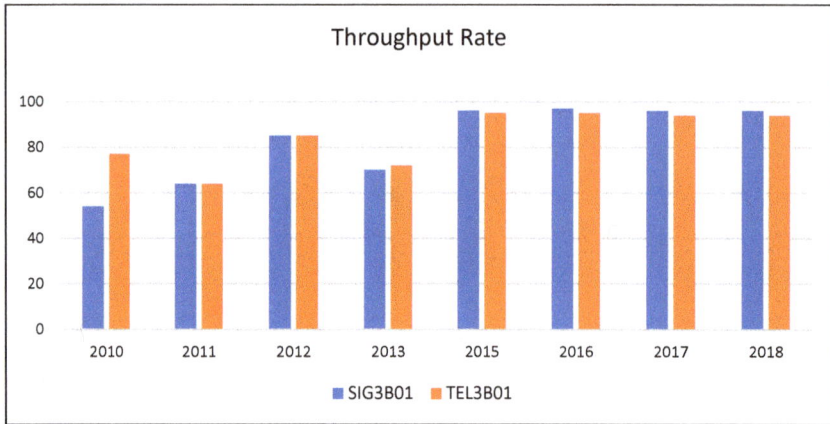

Figure 28. Class pass rate from 2010 to 2018.

5. Conclusions

The improvement of class pass rate was a result of the most important procedure any lecturer should follow; receiving feedback. Feedback in education, whether from students or professionals, helps tremendously in improving the teaching and learning skills of any lecturer. Feedback on the teaching style and skills, teaching philosophy followed by the lecturer, and quality of the course material contribute enormously to the improvement of the quality of the course, the quality of the teaching environment, and thus the improvement of classes with low pass rates.

Funding: This research received no external funding.

Conflicts of Interest: The author declares no conflict of interest.

References

1. Nalla, D.A. Framework for teaching evaluation. In Proceedings of the 2018 IEEE Frontiers in Education Conference (FIE), San Jose, CA, USA, 3–6 October 2018; pp. 1–9.
2. Speaking of Teaching. Using Student Evaluations to Improve Teaching. *Stanf. Univ. Newslett. Teach.* **1997**, *9*, 1–4.
3. Abu Kassim, R.; Johari, J.; Rahim, M. Lecturers' perspective of student online feedback system: A case study. In Proceedings of the 2017 IEEE 9th International Conference on Engineering Education (ICEED), Kanazawa, Japan, 9–10 November 2017; pp. 163–168.
4. Pyasi, S.; Gottipati, S.; Shankararaman, V. SUFAT-An Analytics Tool for Gaining Insights from Student Feedback Comments. In Proceedings of the 2018 IEEE Frontiers in Education Conference (FIE), San Jose, CA, USA, 3–6 October 2018; pp. 1–9.
5. Runniza, S.; Bakri, B.A.; Ling, S.; Julaihi, N.; Liew, C.; Ling, S. Improving low passing rate in mathematics course at higher learning education: Problem identification and strategies towards development of mobile app. In Proceedings of the 2017 International Conference on Computer and Drone Applications (IConDA), Kuching, Malaysia, 9–11 November 2017; pp. 77–81.
6. Sulaiman, F.; Herman, S.H. Enhancing Active Learning through Groupwork Activities in Engineering Tutorials. In Proceedings of the 2nd IEEE International Congress on Engineering Education, Kuala Lumpur, Malaysia, 8–9 December 2010; pp. 106–109.
7. Feisel, L.D. The Role of the Laboratory in Undergraduate Engineering Education. *J. Eng. Educ.* **2005**, *9*, 121–130. [CrossRef]
8. Cojocaru, D.; Popescu, D.; Poboroniuc, M.; Ward, T. Educational policies in European engineering higher education system–Implementation of a survey. In Proceedings of the 2014 IEEE Global Engineering Education Conference (EDUCON), Istanbul, Turkey, 3–5 April 2014; pp. 229–234.

9. Barbara, K. The flipped classroom in engineering education: A survey of the research. In Proceedings of the 2015 International Conference on Interactive Collaborative Learning (ICL), Florence, Italy, 20–24 September 2015; pp. 815–818.

10. Gehringer, E.; Cross, W. A suite of Google services for daily course evaluation. In Proceedings of the 2010 IEEE Frontiers in Education Conference (FIE), Washington, DC, USA, 27–30 October 2010.

11. Wu, B.; Cheng, G. Moodle–The Fingertip Art for Carrying out Distance Education. In Proceedings of the 2009 First International Workshop on Education Technology and Computer Science, Wuhan, China, 7–8 March 2009; pp. 927–929.

12. Andone, D.; Ternauciuc, A.; Vasiu, R. Moodle—Using Open Education Tools for a Higher Education Virtual Campus. In Proceedings of the 2017 IEEE 17th International Conference on Advanced Learning Technologies (ICALT), Timisoara, Romania, 3–7 July 2017; pp. 26–30.

13. Liu, K.; Luo, X.; Xu, Z.; Zhang, J. On Teaching Method and Evaluation Mechanism in Research Teaching. In Proceedings of the 2009 First International Workshop on Education Technology and Computer Science, Wuhan, China, 7–8 March 2009; pp. 580–582.

14. Garry, B.; Qian, L.; Hall, T. Work in progress—Implementing peer observation of teaching. In Proceedings of the 2011 Frontiers in Education Conference (FIE), Rapid City, SD, USA, 12–15 October 2011.

15. Pitterson, N.; Brown, S.; Villanueva, K.; Sitomer, A. Investigating current approaches to assessing teaching evaluation in engineering departments. In Proceedings of the 2016 IEEE Frontiers in Education Conference (FIE), Erie, PA, USA, 12–15 October 2016; pp. 1–8.

16. Yukari, K.; Shotaro, H.; Wataru, T.; Hironori, E.; Masaki, N. FD Commons: E-Teaching Portfolio to Enable an Ubiquitous Peer Reviewing Process. In Proceedings of the 2009 Ninth IEEE International Conference on Advanced Learning Technologies, Riga, Latvia, 15–17 July 2009; pp. 334–336.

17. Wollenberg, B.; Mohan, N. The Importance of Modern Teaching Labs. *IEEE Power Energy Mag.* **2010**, *8*, 44–52. [CrossRef]

18. Pereira, A.; Miller, M. Work in progress—A hands-on ability intervention. In Proceedings of the 2010 IEEE Frontiers in Education Conference (FIE), Washington, DC, USA, 3–6 October 2010.

19. Kampmann, M.; Mottok, J. A "Laboratory" as an approach to foster writing skills at software engineering studies: Learning software engineering is easier when writing courses are directly applied to lecture's content and the problems and examples enrolled in. In Proceedings of the 2018 IEEE Global Engineering Education Conference (EDUCON), Santa Cruz de Tenerife, Canary Islands, Spain, 17–20 April 2018; pp. 900–908.

20. Department of Electrical and Electronic Engineering Science. *Report for the Accreditation Visit of the Engineering Council of South Africa*; University of Johannesburg: Johannesburg, South Africa, 2011.

education sciences

MDPI

Article

Critical Theoretical Frameworks in Engineering Education: An Anti-Deficit and Liberative Approach

Joel Alejandro Mejia [1,*], Renata A. Revelo [2], Idalis Villanueva [3] and Janice Mejia [4]

[1] Department of General Engineering, University of San Diego, 5998 Alcalá Park, San Diego, CA 92110, USA
[2] Department of Electrical and Computer Engineering, University of Illinois—Chicago, 1200 W Harrison St, Chicago, IL 60607, USA; revelo@uic.edu
[3] Department of Engineering Education, Utah State University, Logan, UT 84322, USA; idalis.villanueva@usu.edu
[4] McCormick Office of Undergraduate Engineering, Northwestern University, 633 Clark St, Evanston, IL 60208, USA; j-mejia@northwestern.edu
* Correspondence: jmejia@sandiego.edu

Received: 10 August 2018; Accepted: 19 September 2018; Published: 22 September 2018

Abstract: The field of engineering education has adapted different theoretical frameworks from a wide range of disciplines to explore issues of education, diversity, and inclusion among others. The number of theoretical frameworks that explore these issues using a critical perspective has been increasing in the past few years. In this review of the literature, we present an analysis that draws from Freire's principles of critical andragogy and pedagogy. Using a set of inclusion criteria, we selected 33 research articles that used critical theoretical frameworks as part of our systematic review of the literature. We argue that critical theoretical frameworks are necessary to develop anti-deficit approaches to engineering education research. We show how engineering education research could frame questions and guide research designs using critical theoretical frameworks for the purpose of liberation.

Keywords: critical theoretical frameworks; anti-deficit approach; engineering education research; critical pedagogy

1. Introduction

While critical theoretical frameworks are being used to challenge social practices and belief systems in engineering [1,2], there is a need to dig deeper into the consequences of research whose foci and approach situate underrepresented students as "deficient". Deficit perspectives prevent many underrepresented students and educators from participating in important learning and teaching activities, which further disadvantage students in fields such as engineering [3]. For example, deficit perspectives discourage bilingual children living in high poverty communities from participating in active learning opportunities [4,5].

Ironically, and unfortunately, researchers seeking to understand issues of inclusion, diversity, and retention of underrepresented students could inadvertently ask research questions that focus on the deficits of such populations rather than on their assets [6,7]. Despite deficit perspectives being presented in the literature as lacking empirical validation, research around these beliefs continues to pervade and results in unintended yet dismal consequences on educational practices [6–8].

At the foundation of these deficit perspectives lies the idea that students possess motivational and cognitive deficits. Thus, research that analyzes underrepresented students through deficit-framed questions may perpetuate the idea that these students, particularly students of color and from minoritized groups, have several "needs" [9]. A deficit approach limits the type and forms of interventions that could be tailored to the unique contexts and situated in societies that these students are a part of. To counter this deficit narrative in engineering, it is necessary to pose questions and design studies that provide a better understanding of students' constructed world [9], and how "students of color persist and successfully navigate" [10] (p. 67) different engineering pathways.

Traditional scholarships have been normed by epistemological perspectives that have failed to examine structures of domination and oppression in educational settings [11]. As such, there is a need for more diversity in the methods and theoretical frameworks used in engineering education to frame and design research studies that challenge deficit models and center on the assets of underrepresented students rather than their deficits [12]. This study presents ways in which critical theoretical frameworks can be used in fields like engineering education following Freire's principles of critical andragogy and pedagogy [13]. Engineering education was selected for this study due to its recent inception into the research realm [14–18] and due to its normative, hegemonic (primarily composed of white, male, and middle class), and reliance upon meritocratic ideologies [15,16,19–22]. In addition, the study outlines the characteristics of anti-deficit scholarship, and describes the implications of connecting reflection, practice, and research to achieve transformative changes in engineering education.

2. Positionality

Educators must reflect into their own biases and limitations and disengage from framing questions that may potentially lead to unintentional promotion of deficit perspectives. Awareness through reflection is the first step to engage in both pedagogical and andragogical research practices that counter such perspectives [23,24]. Critical reflection, in particular, is crucial because it can uncover the power dynamics that exist in engineering education, and can help to challenge hegemonic assumptions about students [25]. Thus, researchers must frame questions in terms of power, privilege, and oppression, while engaging in critical reflection [1,26].

The research team encompasses minoritized populations in engineering and engineering education research whose asset-based approaches have helped them excel in engineering. As being placed in positions of power and authority through their existing roles as university faculty at research and teaching institutions, it is the authors' belief that bringing to light this analysis will further encourage future populations of underrepresented and minoritized populations in engineering, via newly informed educational practices, to succeed and persist in the field.

The research team also posits that communities of color and other minoritized groups (e.g., LGBTQ) deserve to be validated and acknowledged by the general and engineering education community. These communities have a wealth of knowledge, skills, and practices that are very rich and powerful [27]. Their meaning-making practices cannot and should not be silenced, sanctioned, nor neglected [9,27]. A different worldview of engineering is not a "deficient" interpretation of engineering [27], but a manifestation of the cultural, historical, and social richness of communities of minoritized groups (including but not limited to people of color). Thus, asset-based approaches create the bridges necessary to merge both the formal and informal spaces while acknowledging the lived realities and embodied knowledge of students of color and other minoritized groups in engineering.

3. Theoretical Framework

3.1. Concientização and Praxis

Freire argued that reflection is necessary because it seeks to overcome the alienating and dehumanizing situation of many individuals [28,29]. A lack of reflection prevents individuals from forming the cognitive and motivational tools needed to liberate themselves from the conditioning and historical factors that hinder their development [13,28,30]. Transformative changes cannot be achieved unless there is a combination of action built upon reflection, or concientização [1,13], and a theory merged with action, or *praxis* [1,13].

Concientização involves three phases of what the individual goes through to achieve liberation: (1) Magical, (2) naïve, and (3) critical. The magical phase is characterized by being in a state of impotence where the individual is unable to do anything about their situation, which was created by the system. As such, the individual is controlled by outside forces (the system or social structure that they are part of) that are simply viewed as causality to the individual's situation [23]. The naïve phase is characterized by a meek understanding of one's situation within the system but with the internalized certainty that one is not capable of changing it. The individual accepts that there are aspects of their life within their reach and others that are not within their reach. The critical phase involves uncovering factors that make individuals different, and due to this distinction, they are able to understand the ways in which the system can be unfair. During the critical phase, individuals realize that the system can be transformed by removing oppressive ideologies and reclaiming power for the benefit of the community [13,28].

For Freire, concientização was always inseparable from liberation. The liberation process is characterized by dialogue [30] and concientização is the ability to hold the most critical possible view of reality [23,28]. In the model presented in Figure 1, liberation is reached when there is a combination of theory developed from reflection (e.g., scholarship), action guided by theory (e.g., praxis), and action based on reflection (e.g., concientização). Liberation is also believed to be characterized by radically transforming praxis, and should be understood as a pedagogical method of liberation of oppressed people, although it can be generalized to all types of education and to all types of society, poor or developed [23]. Thus, theory development that is done through reflection leads to scholarship; action that takes into consideration the theory developed can evolve into praxis; and action based on reflection develops into concientização.

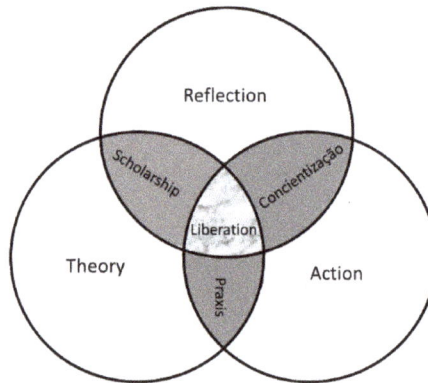

Figure 1. Proposed critical consciousness approach by the authors; model was adapted and expanded upon Freire's principles of critical pedagogy [13].

3.2. Critical Theoretical Frameworks

According to Horkheimer [31], there is a distinction between traditional and critical theory. Traditional theory seeks to only understand or describe society, while critical theory seeks to critique and change society. Critical theory recognizes the complexity of social processes and its main task is "to reflect upon the structures from which social realities and the theories that seek to explain it are constructed" [32] (p. 139). Critical theory seeks not only to critique society but also to provide the foundation to transform society as a whole [32]. From its origins in the Frankfurt School and its focus on a criticism of modern social structures [32], critical theory has continued to grow and contribute to the bases for inquiry in other fields such as sociology and education [33–35], pedagogy [13,28,30,36–40], andragogy [41], and other areas including feminism, law, and social sciences [42–51].

Critical theories not only describe and critique complex social constructions, they also explore the circumstances that lead to oppression. For instance, Critical Race Theory (CRT) emerged from legal scholarship to provide an overview of the permeation of racism through the legal system [48]. Eventually, scholarship in education integrated CRT to examine and challenge the traditional paradigms that exist in the educational system [37], and analyze how inequalities in schools are the result of a racialized society [39]. CRT has been used as a framework to describe the experiences of students of color in hegemonic spaces, to counter dominant paradigms and narratives, and to better understand the role of agency and empowerment of the oppressed [11,52,53].

Thus, critical theories have been used to provide new perspectives and advocate for an approach that is not primarily positivist or that uses methods to classify the social world in an objective way with causal connection [53–55]. These theories illustrate the ways in which context, gender, culture, society, and other factors can be studied through a critical lens, in order to achieve equity. The impact of critical theories in engineering education could potentially address the problems of underrepresentation in the field by challenging its uninterrogated and institutionalized norms. Nonetheless, using these critical lenses to understand the experiences of underrepresented students in engineering could be detrimental if praxis (the fusion of theory and action) and concientização (action based on reflection) are not achieved [13]. Thus, understanding how to appropriately use critical theoretical frameworks could help scholars and educators analyze the engineering climate, its impact on minoritized populations, guide future research, and provide an opportunity to further improve the ways in which engineering can become more inclusive instead of superficially diverse.

3.3. Anti-Deficit Framework

Harper [10] provided a framework that redefines how research questions in science, technology, engineering, and math (STEM) research are framed. He argued that traditional scholarship encompasses a considerable amount of emphasis on "understanding how [underrepresented students] managed to acquire various forms of capital they did not possess upon entry" to different STEM fields [10]. Harper sought to invert the logic of traditional research studies in STEM education by asking questions such as "what do students of color have?", rather than "what do students of color need?". In this way, asset-based approaches promote learning in meaningful and relevant ways while challenging deficit-thinking models.

Aligned with Harper's views, the authors of this paper posit that critical theoretical frameworks, particularly in engineering education research, should explore how reflection, theory and action merge to not only achieve concientização but also praxis. Furthermore, through an anti-deficit approach, educators, students, researchers and their intended participant populations can experience liberation from oppressive forces and their scholarship can serve as a tool to promote this freedom.

As illustrated in Figure 1, the integration of the three concepts of critical pedagogy (theory, reflection, and action) presented by Freire [13] can help conceptualize a new approach to fields like engineering education research. This model was built upon Freire's conceptualization and synthesis of critical pedagogy to show visually how scholarly work can be driven by critical theoretical perspectives. At the intersection of theory and reflection is Scholarship—a state where the scholar has a deep understanding of their reflexivity as a researcher and educator and uses critical theory to inform and conduct research. At the intersection of reflection and action is Concientização—a state where the scholar has a deep understanding of their reflexivity as a researcher and educator, and enacts it not only in research but also in other forms of scholarly work (e.g., advocacy, programs, teaching, service). At the intersection of theory and action is Praxis—a state where the scholar uses critical theory to inform their scholarly work (e.g., research, advocacy, programs, teaching, service). At the intersection of theory, reflection, and action is Liberation—a state where the scholar has reached critical consciousness in all aspects (i.e., theory, reflection, and action) and enacts critical consciousness in all aspects of their work.

The following study describes how the authors investigated the ways in which engineering education research that has used critical theoretical frameworks in their design has yet to detach itself from research questions framed in a deficit-oriented manner. We focused on studies that utilized critical theoretical frameworks since the foundation of this type of research is to critically analyze how oppressive systems are created and to provide empowerment and liberation for those who are marginalized [13,28,36,52,53].

4. Methodology

4.1. Research Design

This study is built upon larger and more comprehensive study conducted by the research team [1,26,56]. The research team applied qualitative approaches, specifically Critical Discourse Analysis [57], to the analysis of the literature, and applied an interpretive philosophical lens to the findings [58,59]. As indicated by Fairclough [57], Critical Discourse Analysis "takes particular interest in the relation between language and power" (p. 2). Part of using language in particular ways through scholarship is to produce representations and social process and practices that shape specific discourses [57]. Thus, Critical Discourse Analysis was used to assess how engineering education, as an emerging field, has embraced the adoption and application of critical theoretical frameworks. In addition, the intent of the analysis was to identify how the use of critical theoretical frameworks guided the research directed primarily at minoritized populations.

4.2. Review Questions

The central review questions that guided this study included:

1. What are the common types of critical theoretical frameworks used to study underrepresented populations in engineering education?
2. How are these critical theoretical frameworks used in within research methodologies for these engineering populations?

4.3. Systemic Literature Review

A systematic literature review was conducted as recommended by Khan and colleagues [60] where the authors derived from our formulated review questions, identified relevant studies, selected studies that fit the inclusion criteria, appraised the quality of the research studies, and summarized the evidence by use of an explicit methodology. Based upon our research questions, relevant studies were identified through the following databases: ERIC, IEEE Xplore, Journal of Higher Education, Journal of Engineering Education, ASEE PEER, Journal of Women and Minorities in Science and Engineering, and the Journal of STEM Education. The descriptors "critical theory", "underrepresented minority", "critical race theory", "feminism", "conciencia", and "intersectionality" were used to locate primary sources. These descriptors were also used in conjunction with other descriptors such as "underrepresented populations", "Latino", "Hispanic", "African American", "Native American", and "women" as these are all underrepresented populations in engineering [61].

Several articles were identified as potential sources of information, but to assess their quality, only articles that met the following inclusion criteria were reviewed: (1) Published between 2005 and 2016; (2) listed critical theoretical frameworks as one of the lenses for analysis; and (3) investigated K-16 academic engineering education. The papers were divided into the types of critical theoretical frameworks listed in the different manuscripts identified. In total, there were 33 articles reviewed that represented a wide variety of critical theoretical frameworks. Of those 33 articles, only 28 clearly identified the type of critical theoretical framework used. The other five articles simply indicated that the frameworks used or developed were "critical". Each article was reviewed by at least one of the four authors in detail using an agreed-upon coding sheet.

To summarize the evidence found in these identified articles, a coding sheet was developed based on the characteristics significant to each study evaluated. These categories on the coding sheet included identifying the purpose of the study, the research questions, the methods used, the type of data collected, the population involved in the study, and relevant findings. Additionally, we used our adapted Freirian critical consciousness model (Figure 1) to understand the ways in which the critical theoretical frameworks were used in these publications. After reviewing the articles, we synthesized the preliminary findings and patterns each author saw in their respective notes. The lead author reviewed the notes and preliminary findings to guide the final review.

After the articles were analyzed, the authors developed a representative table of different ways in which the articles framed the research questions. These representative examples were not taken verbatim, but rather synthesized to illustrate how critical research in engineering education can reframe deficit-oriented to anti-deficit questions, as indicated by Harper [10], when guided by critical theory. In addition, we identified Freire's principles of critical pedagogy [13] that were emphasized in the studies to describe how critical theoretical frameworks in engineering education are primarily enacted. In the Results section, we illustrate how we reframed representative research questions posed in the engineering education research studies reviewed and the intersecting principles (theory, reflection, and action) highlighted by the studies analyzed.

4.4. Limitations

While this study was conducted on work related to engineering education, it is possible that the authors may have omitted research studies on "STEM" that may have included engineering populations. Furthermore, it is recognized that by selecting publications from 2005 to 2016, we may have omitted earlier studies in engineering. However, the focus of this work was to explore the state-of-the-art of these types of studies in engineering education. Finally, we want to acknowledge that some of the studies reviewed used more than one critical theoretical framework (e.g., Community Cultural Wealth and Funds of Knowledge). However, within our inclusion criteria, we focused on studies that used at least one critical framework and did not analyze the impact of those that may have used a combination of these frameworks.

5. Results

The articles identified in the systematic literature review illustrates the growing number of studies that employed a critical theoretical framework by the engineering education research community to explore the histories and experiences of underrepresented populations in engineering. As shown in Table 1, most of the studies incorporated feminist theory (and its variants of feminist thought such as Womanism and Mujerismo) or CRT to analyze the social dynamics in engineering. Other common types of critical theoretical frameworks included intersectionality, community cultural wealth, funds of knowledge, and Bourdieuian frameworks.

The variety of critical theoretical frameworks indicated the openness and effort from the engineering education research community to integrate theoretical lenses to challenge the status quo. However, in terms of praxis, several studies did not combine theory and action while engaging with minoritized groups to ask deeper questions related to power, oppression, and normative practices. Asking critical questions that challenge systems of oppression comprises one of the requirements of critical theoretical frameworks for engineering education. For example, the majority of the papers provided insights and implications for the work but seldom was there evidence of researchers taking actions to fight alongside the participants. Freire argues for the need for revolutionary educators to "fight alongside the people" and not just to 'win the people over' [13] (pp. 94–95).

In answering the review question on how critical frameworks were being used, the authors noted the language used in the research questions and throughout the introductory sections to describe the experiences of these underrepresented groups in engineering. For instance, many of these studies focused on describing a "deficiency" first (e.g., the lack of language proficiency or immediate support networks), rather than a characteristic that these populations could "voice" to challenge deficit models, or describing the normative bases and structure for social/educational inequity. The results of some of these studies indicated that students of color "lacked" family support networks or role models, or disregarded the fluidity of identity formation [1,26]. In addition, the language used in many of the research questions were also framed with a deficit perspective. Representative paraphrased examples of these research questions are included in Table 2 along with examples of ways in which these research questions could be reframed using an anti-deficit and critical theory-guided approach. While Table 2 provides only representative paraphrased examples of research questions, the authors noted that all of the articles reviewed asked questions that were driven by Scholarship—the intersection of theory and reflection. The lack of Praxis or Concientização in these studies suggests a need to explore these further in engineering education research. Few, if any, papers provided anti-deficit research questions or frameworks.

Table 1. Frequency of critical theoretical frameworks used in engineering education research from 2005 to 2016 addressing underrepresented minorities.

Critical Theoretical Framework	Population Addressed	Frequency of Studies
Critical Race Theory	Asian Americans, African Americans, Latinxs	5
Feminist Theory (including the variants mujerismo and womanism)	Women	5
Intersectionality Theory	Women	5
Community Cultural Wealth	Latinxs, African Americans	4
Funds of Knowledge	Latinxs, First-generation students	3
Identity Theory	First-generation students	2
Burdieuian Frameworks (e.g., social capital, cultural capital, habitus, socialization)	African Americans, Latinxs, Asian Americans, White women	2
Critical Agency	Dominant and non-dominant groups in engineering	2
Not clearly identified or defined	Dominant and non-dominant groups in engineering	5

Table 2. Representative paraphrased examples of research questions analyzed in the systemic review of critical engineering education research.

Deficit-Oriented Questions	Anti-Deficit Reframing	Critical Theory Guided Questions
To what extent do Black engineering students participate in engineering extracurricular activities?	What stimulates Black engineering students to participate in engineering extracurricular activities?	How do engineering extracurricular activities promote Black engineering student participation?
Why do Latinx students leave the engineering pipeline?	What compels Latinx students to persist in engineering despite the institutional challenges?	What institutional challenges prevent Latinx students to persist in engineering?
Why are Native American students unprepared for engineering courses?	How do Native American students overcome educational disadvantages?	How are institutions responsive to varied levels of educational preparation for Native American students?
Why do students of color not pursue graduate degrees in engineering?	What are the typical pathways toward doctoral degrees for students of color?	What aspects of graduate education in engineering reinforce inequality for students of color?

The systemic review indicated that critical theoretical frameworks are being used in engineering education primarily for the sake of theory development for either research or practice. For example, only one of the articles reviewed used principles of participatory action research methods where participants were partakers of actionable outcomes with the researchers to pursue their study. Based on our model, such an approach for research was the only example we found in our review of Praxis. In general, the studies did not take a critical stance on how engineering knowledge is constructed, who participates in engineering, and who decides who becomes an engineer. In terms of critical pedagogy, few studies questioned how to empower students of color (e.g., concientização) or considered taking action and working alongside the students (e.g., praxis) to de-colonize and re-inhabit their spaces, including all of these different domains that students of color inhabit.

The rationale or motivation to study minoritized populations in engineering was also reviewed to determine how critical theoretical frameworks were being adapted and applied when working with these populations. The analysis indicated that most studies focused on a deficiency-driven perspective. This type of research framing provides a narrow view of minoritized populations that can eventually become part of a larger dominant discourse in engineering. For example, there was a predominant notion that students (particularly low-income, underrepresented students) fail in school because such students and their families experience deficiencies that obstruct the learning process (e.g., lack of motivation and inadequate home socialization). Unfortunately, some of these studies did not interrogate power structures and epistemological frameworks that perpetuate this narrative.

One emergent finding to note was that few studies considered how different identities intersect. For example, one of the studies was aimed to study feminist theories and intersectionality in engineering, but upon closer examination, the population did not include literature or rationale for not including international and national non-English language women participants. In several studies, the term Hispanic was mentioned as the population of study, yet there was no distinction between this definition and Latin@/Chican@ nor were there references about the community, demographics, language, culture, etc., that would precisely "de-cluster" these groups [26]. The same applied for engineering education research on Native American and African American populations whose demographics, origins, and sub-cultures were not considered. Neither example provided several implications for the study but failed to suggest strategies to challenge power structures to dismantle oppression of people of color in engineering. The latter finding was seen across all other manuscripts reviewed.

Finally, none of the studies reviewed paid significant attention to the historical contexts of the populations studied. The context provided was limited to the sample of the population studied (e.g., traditional age, women, and residential) or institution studied (e.g., private, predominantly white).

6. Discussion

The systematic literature review demonstrated that, although research in engineering education is increasingly adopting critical theoretical frameworks, the intended outcomes of using critical frameworks, such as Concientização and Praxis [31,62], are not being addressed, and as a result, the research does not achieve Liberation. Most of the studies analyzed in this systematic literature review focused on a combination of theory and reflection to produce Scholarship. It is important for engineering education researchers to take into consideration all principles of Freire's critical pedagogy [13] in order to achieve Liberation, which is the state at where a scholar has reached critical consciousness. This action requires integrating all elements of the proposed model (Figure 1).

The findings positioned the authors of this paper to question if selection of critical frameworks for engineering is effectively describing the lived experiences of marginalized groups or achieving the outcomes established by these frameworks. As more and more studies in engineering begin to focus on other dimensions of underrepresentation such as language, immigration, ethnicity, culture, identity, phenotype, sexuality, among others, it will be important for educators/researchers to have a targeted lens when exploring these complex yet important phenomena.

As future uses of critical frameworks in engineering continue, it will be important to consider more purposeful sampling of these underrepresented and minoritized groups. Limiting sampling methods and approaches in critical analysis work could be detrimental to the goal of praxis and concientização, and risk the unintended invalidation or belittling of cultures, languages, and experiences, and never reach the liberation the research intends to achieve.

7. Conclusions

While a robust corpus of literature exists on engineering education research that utilizes critical theoretical frameworks, deficit models still persist in how research is framed, thus normalizing the idea that students of color have several "needs". Re-imagining engineering education from an asset-based approach has a strong propensity to develop a knowledgeable citizenry who understands the importance and value of our human constructed world, while validating and acknowledging the contributions of people of color and minoritized groups to engineering. As a field dominated by hegemonic practices and norms, engineering is a field that greatly needs critical perspectives that could help deconstruct dominant discourses—the combination of language, tools, actions, interactions, technologies, processes, beliefs, and values [63]. The Critical Discourse Analysis approach used in this study highlights the importance of dismantling language and power in traditional scholarship in engineering education. Particularly, it is important to analyze how engineering education may reproduce specific discourses that perpetuate deficit models through deficit-oriented questions and practices. For example, some of the studies included in this review chose to use the word "critical" as a way to describe individual reflexive processes and differences between individuals (primarily people of color and minoritized groups), rather than looking at systems of oppression. There is an underlying assumption that "critical" means being descriptive and reflexive of specific phenomena rather than working through positionality to challenge systemic and institutionalized forms of oppression.

8. Implications/Recommendations

After a careful analysis, the authors opted to take action on this paper by developing a series of guiding questions that can help researchers combine all principles of critical pedagogy to achieve liberation for minoritized and underrepresented groups in fields like engineering. As listed in Table 3, sample questions were re-written in a way that considers how praxis and concientização can be accounted for in addition to scholarship for the goal of liberation. It is important to note that throughout the process, some type of member checking needs to take place so that the researchers share their understanding of the results with the participants in an effort to enhance the credibility of findings and trustworthiness [64]. It is important to reflect on each other's understanding of the phenomena being studied, revise the results, take action, and co-create theory together to achieve liberation. These guiding questions are not intended to be prescriptive, but rather to be considered by researchers who are interested in using critical theoretical frameworks for the study of minoritized populations in engineering.

Table 3. Guiding questions for researchers using the adapted model of Freire's principles of critical pedagogy from Figure 1.

Freire's Principles of Critical Pedagogy	Theory	Action	Reflection
Scholarship	Is this theory critical and am I considering the political, cultural and historical factors that play a role into the research?	In what ways is my research and my relationship with the participants ensuing that a liberating action will occur?	What is my positionality?
Praxis	Are the theories that I am trying to explore achieving the intended goal?	How do I ensure that my research results can be easily translated into practice?	How am I reflecting upon my role as a researcher in the context of the phenomenon/population I am trying to explore?
Concientização	Does the theory used assume a deficit or anti-deficit approach?	What are my assumptions about the community and the phenomenon?	In what ways was I mistaken about the population or the phenomenon I explored?
Liberation (e.g., for participants)	How can I make sure the theory development in my work is liberative and co-created with participants?	In what ways am I allowing participants to take action alongside me in order to achieve liberation from the obstacles that prevent action from occurring?	In what ways am I allowing for participants to reflect with theme about the research findings and to co-construct these narratives together?

Furthermore, for researchers interested in conducting work in fields like engineering, reflection, theory, and action should not be seen as separate from each other. Another element to consider is that critical researchers do not only describe an event or experiences; they ask questions of power, privilege, and oppression. Engineering is situated against historical underrepresentation of people of color in STEM [65]. If engineering forms from culture and practices, it is a challenge for students of color and other minoritized groups who want to become part of the dominant culture because of what they have to sacrifice, especially when dominant paradigms and deficit models are disseminated through the work of engineering education researchers.

In addition, we cannot separate out the epistemic knowledge from ontological aspects of identities and the realities of minoritized and underrepresented students' everyday lives. For instance, identity work is longitudinal and requires recognition from others. Students start to perform as engineers throughout their educational experiences; however, those experiences are laminated longitudinally. Instead of focusing on the actual student identity or experiences, how can we begin to change the environments? How do we create the spaces in which students are allowed to participate, and redefine and reimagine the notions of what participation means?

Finally, although it was unclear what the researchers' intentions were in pursuing critical research in engineering education, we maintain that there are changes that need to be made for future studies. It is important to engage in critical reflection while the research is being conducted. If the idea is to incorporate critical theoretical frameworks in engineering education research, it is imperative that the researchers include methods that incorporate strategies to "fight alongside" [13] (pp. 94–95) students of color. Borrowing from Freire's work [13], we use the phrase "fight alongside" purposefully to propose a stance where researchers/educators/scholars take a liberated perspective into all of their work. Using such a stance implicates that researchers take an approach that is not just theory-based, but also incorporates reflection and action. We note that from our review, we found most studies were limited to Scholarship (i.e., intersection of theory and reflection) and lacked the action tenet. Additionally, there is no accountability on the methods in which praxis and concientização are achieved. We recommend that research that integrates critical theoretical frameworks include sections on methods, positionality, and reflection that allow the readers to learn more about how to create and sustain transformative approaches.

Author Contributions: Conceptualization, J.A.M., R.A.R. and I.V.; Methodology, J.A.M., R.A.R. and I.V.; Formal Analysis, J.A.M., R.A.R., I.V., and J.M.; Resources, J.A.M., R.A.R., I.V., and J.M.; Conceptualization of Freirian Model, J.A.M., R.A.R. and I.V.; Conceptualization of Tables 2 and 3, J.A.M., R.A.R., I.V., and J.M.; Writing-Original Draft Preparation, J.A.M.; Writing-Review & Editing, J.A.M., R.A.R., I.V., and J.M.

Funding: This material is based upon work supported by the National Science Foundation (NSF) under Grant No. DRL-1644976 and EEC-1653140. Any opinions, findings, and conclusions or recommendations expressed in this material does not necessarily reflect those of NSF.

Conflicts of Interest: The authors declare no conflict of interest.

References

1. Mejia, J.A.; Revelo, R.A.; Villanueva, I. The "fibonacci sequence" of critical theoretical frameworks: Breaking the code of engineering education research with underrepresented populations. In Proceedings of the ASEE Annual Conference and Exposition, Columbus, OH, USA, 25–28 June 2017.
2. Pawley, A.; Mejia, J.A.; Revelo, R.A. Translating theory on color-blind racism to an engineering education context: Illustrations from the field of engineering education research with underrepresented populations. In Proceedings of the ASEE Annual Conference and Exposition, Salt Lake City, UT, USA, 24–27 June 2018.
3. Esquinca, A. Socializing pre-service teachers into mathematical discourse: The interplay between biliteracy and multimodality. *Multiling. Educ.* **2012**, *2*, 1–20. [CrossRef]
4. Adair, J.K.; Colegrove, K.S.-S.; McManus, M.E. How the word gap argument negatively impacts young children of Latinx immigrants' conceptualizations of learning. *Harv. Educ. Rev.* **2017**, *87*, 309–334. [CrossRef]
5. Colegrove, K.S.-S.; Adair, J.K. Countering deficit thinking: Agency, capabilities and the early learning experiences of children of Latina/o immigrants. *Cont. Issues Early Child.* **2014**, *15*, 122–135. [CrossRef]

6. Valencia, R.R.; Solórzano, D.G. *Contemporary Deficit Thinking*; Falmer: New York, NY, USA, 1997; pp. 160–210.
7. Valenzuela, A. *Subtractive Schooling: US-Mexican Youth and the Politics of Caring*; State University of New York Press: Albany, NY, USA, 2010.
8. Valencia, R.R. The Mexican American struggle for equal educational opportunity in Mendez v. Westminster: Helping to pave the way for Brown v. Board of Education. *Teach. Coll. Rec.* **2005**, *107*, 389–423. [CrossRef]
9. Mejia, J.A.; Wilson-Lopez, A.; Robledo, A.L.; Revelo, R.A. Nepantleros and nepantleras: How Latinx adolescents participate in social change in engineering. In Proceedings of the ASEE Annual Conference and Exposition, Columbus, OH, USA, 25–28 June 2017.
10. Harper, S.R. An anti-deficit achievement framework for research on students of color in STEM. *New Dir. Inst. Res.* **2010**, *148*, 63–74. [CrossRef]
11. Delgado Bernal, D. Using a Chicana feminist epistemology in educational research. *Harv. Educ. Rev.* **1998**, *68*, 555–583. [CrossRef]
12. Mein, E.; Esquinca, A.; Monarrez, A.; Saldaña, C. Building a pathway to engineering: The influence of family and teachers among Mexican-origin undergraduate engineering students. *J. Hispan. High. Educ.* **2018**, *1*, 1–15. [CrossRef]
13. Freire, P. *Pedagogy of the Oppressed*; Bloomsbury: New York, NY, USA, 2003.
14. Jesiek, B.K.; Newswander, L.K.; Borrego, M. Engineering education research: Discipline, community, or field? *J. Eng. Educ.* **2009**, *98*, 39–52. [CrossRef]
15. Davis, M. *Engineering as Profession: Some Methodological Problems in Its Study*; Springer: New York, NY, USA, 2015; pp. 65–79.
16. Lohmann, J.; Froyd, F. Chronological and ontological development of engineering education as a field of scientific inquiry. Presented at the Second Meeting of the Committee on the Status, Contributions, and Future Directions of Discipline-Based Education Research, Washington, DC, USA, 2010; Available online: https://sites.nationalacademies.org/cs/groups/dbassesite/documents/webpage/dbasse_072587.pdf (accessed on 22 September 2018).
17. Borrego, M. Development of engineering education as a rigorous discipline: A study of the publication patterns of four coalitions. *J. Eng. Educ.* **2007**, *96*, 5–18. [CrossRef]
18. Koro-Ljungberg, M.; Douglas, E.P. State of qualitative research in engineering education: Meta-analysis of jee articles, 2005–2006. *J. Eng. Educ.* **2008**, *97*, 163–175. [CrossRef]
19. Downey, G.L.; Lucena, J.C. Knowledge and professional identity in engineering: Code-switching and the metrics of progress. *Hist. Technol.* **2004**, *20*, 393–420. [CrossRef]
20. Baillie, C.; Pawley, A.; Riley, D.M. *Engineering and Social Justice: In the University and Beyond*; Purdue University Press: West Lafayette, IN, USA, 2012.
21. Jolly, L.; Radcliffe, D. Reflexivity and hegemony: Changing engineers. In Proceedings of the 23rd HERDSA Conference, Toowoomba, Australia, 2–5 July 2000; pp. 357–365.
22. Riley, D. *Engineering and Social Justice: Synthesis Lectures on Engineers, Technology, and Society*; Morgan and Claypool: San Rafael, CA, USA, 2008; pp. 1–152.
23. Lawrence, L.C. La concientización de paulo freire. *Historia de la Educación Colombiana* **2008**, *11*, 51–72.
24. Melnyk, R.; Novoselich, B.J. The role of andragogy in mechanical engineering education. In Proceedings of the ASEE Annual Conference and Exposition, Columbus, OH, USA, 25–28 June 2017.
25. Brookfield, S. The concept of critical reflection: Promises and contradictions. *Eur. J. Soc. Work* **2009**, *12*, 293–304. [CrossRef]
26. Revelo, R.A.; Mejia, J.A.; Villanueva, I. Who are we? Beyond monolithic perspectives of Latinxs in engineering. In Proceedings of the ASEE Annual Conference and Exposition, Columbus, OH, USA, 25–28 June 2017.
27. Mejia, J.A.; Pulido, A. Fregados pero no jodidos: A case study of Latinx rasquachismo. In Proceedings of the ASEE Annual Conference and Exposition, Salt Lake City, UT, USA, 24–27 June 2018.
28. Freire, P. *The Politics of Education: Culture, Power, and Liberation*; Bergin and Garvey: South Hadley, MA, USA, 1985.
29. Giroux, H.A. Rethinking education as the practice of freedom: Paulo Freire and the promise of critical pedagogy. *Policy Futures Educ.* **2010**, *8*, 715–721. [CrossRef]
30. Freire, P. Reading the world and reading the word: An interview with Paulo Freire. *Lang. Arts* **1985**, *62*, 15–21.

31. Horkheimer, M. *Traditional and Critical Theory: Selected Essays*; Herder and Herder: New York, NY, USA, 1976; pp. 188–243.
32. González, F.; Moskowitz, A.; Castro-Gómez, S. Traditional vs. Critical cultural theory. *Cult. Crit.* **2001**, *49*, 139–154.
33. Bourdieu, P. The forms of capital. *Cult. Theory Anthol.* **2011**, *1*, 81–93.
34. Bourdieu, P.; Passeron, J.-C. *Reproduction in Education, Society and Culture*; Sage: Beverly Hills, CA, USA, 1990.
35. Bourdieu, P.; Wacquant, L.J. *An Invitation to Reflexive Sociology*; University of Chicago Press: Chicago, IL, USA, 1992.
36. Shor, I.; Freire, P. *A Pedagogy for Liberation: Dialogues on Transforming Education*; Bergin and Garvey: South Hadley, MA, USA, 1987.
37. Ladson-Billings, G. Toward a theory of culturally relevant pedagogy. *Am. Educ. Res. J.* **1995**, *32*, 465–491. [CrossRef]
38. Ladson-Billings, G. Just what is critical race theory and what's it doing in a nice field like education. *Qual. Stud. Educ.* **1996**, *11*, 7–24. [CrossRef]
39. Ladson-Billings, G.; Tate, W.F. Toward a critical race theory of education. *Teach. Coll. Rec.* **1995**, *97*, 47–68.
40. Ladson-Billings, G.; Tate, W.F. *Education Research in the Public Interest: Social Justice, Action, and Policy*; Teachers College Press: New York, NY, USA, 2006.
41. Wilson, A.L.; Kiely, R.C. Towards a critical theory of adult learning/education: Transformational theory and beyond. In Proceedings of the Adult Education Research Conference, Raleigh, NC, USA, 24–26 May 2002.
42. Foucault, M. The history of sexuality: An introduction, volume i. In *Trans. Robert Hurley*; Vintage: New York, NY, USA, 1990.
43. Crenshaw, K. *Critical Race Theory: The Key Writings that Formed the Movement*; The New Press: New York, NY, USA, 1995.
44. Crenshaw, K. Demarginalizing the intersection of race and sex: A black feminist critique of antidiscrimination doctrine, feminist theory and antiracist politics. *U. Chi. Legal F.* **1989**, *1*, 139–167.
45. Crenshaw, K.W. Toward a race-conscious pedagogy in legal education. *Nat'l Black LJ* **1988**, *11*, 1–14.
46. Delgado, R. When a story is just a story: Does voice really matter? *Va. Law Rev.* **1990**, *76*, 95–111. [CrossRef]
47. Delgado, R.; Stefancic, J. Critical race theory: An annotated bibliography. *Va. Law Rev.* **1993**, *79*, 461–516. [CrossRef]
48. Delgado, R.; Stefancic, J. *Critical Race Theory: An Introduction*; NYU Press: New York, NY, USA, 2017.
49. Delgado-Bernal, D. Critical race theory, latino critical theory, and critical raced-gendered epistemologies: Recognizing students of color as holders and creators of knowledge. *Qual. Inq.* **2002**, *8*, 105–126. [CrossRef]
50. Delgado-Bernal, D.; Villalpando, O. An apartheid of knowledge in academia: The struggle over the "legitimate" knowledge of faculty of color. *Equity Excell. Educ.* **2002**, *35*, 169–180. [CrossRef]
51. Bell, D.A. Serving two masters: Integration ideals and client interests in school desegregation litigation. *Yale Law J.* **1976**, *85*, 470–516. [CrossRef]
52. Solorzano, D.G.; Yosso, T.J. Critical race and latcrit theory and method: Counter-storytelling. *Int. J. Qual. Stud. Educ.* **2001**, *14*, 471–495. [CrossRef]
53. Solórzano, D.G.; Yosso, T.J. Critical race methodology: Counter-storytelling as an analytical framework for education research. *Qual. Inq.* **2002**, *8*, 23–44. [CrossRef]
54. Giroux, H. *Pedagogy and the Politics of Hope: Theory, Culture, and Schooling: A Critical Reader*; Routledge: Abingdon-on-Thames, UK, 2018.
55. Giroux, H. Theories of reproduction and resistance in the new sociology of education: A critical analysis. *Harv. Educ. Rev.* **1983**, *53*, 257–293. [CrossRef]
56. Villanueva, I.; Gelles, L.; Di Stefano, M.; Smith, B.; Tull, R.; Lord, S.; Benson, L.; Hunt, A.; Riley, D. What does hidden curriculum in engineering look like ad how can it be explored? In Proceedings of the American Society of Engineering Education Annual Conference and Exposition, Salt Lake City, UT, USA, 24–27 June 2018.
57. Fairclough, N. *Critical Discourse Analysis: The Critical Study of Language*; Routledge: New York, NY, USA, 2013.
58. Creswell, J.W.; Creswell, J.D. *Research Design: Qualitative, Quantitative, and Mixed Methods Approaches*; Sage Publications: Thousand Oaks, CA, USA, 2017.
59. Denzin, N.K.; Lincoln, Y.S. *The Sage Handbook of Qualitative Research*; Sage: Beverly Hills, CA, USA, 2011.
60. Khan, K.S.; Kunz, R.; Kleijnen, J.; Antes, G. Five steps to conducting a systematic review. *J. R. Soc. Med.* **2003**, *96*, 118–121. [CrossRef] [PubMed]

61.   President's Council of Advisors on Science and Technology. *Engage to Excel: Producing One Million Additional College Graduates with Degrees in Science, Technology, Engineering, and Mathematics*; President's Council of Advisors on Science and Technology: Washington, DC, USA, 2012.

62.   Horkheimer, M. *Critical Theory*; Continuum: New York, NY, USA, 1982.

63.   Gee, J.P. *Social Linguistics and Literacies: Ideology in Discourses*, 5th ed.; Routledge: New York, NY, USA, 2015.

64.   Thomas, D.R. A general inductive approach for analyzing qualitative evaluation data. *Am. J. Eval.* **2006**, *27*, 237–246. [CrossRef]

65.   Landivar, L.C. *Disparities in Stem Employment by Sex, Race, and Hispanic Origin*; U.S. Census Bureau: Washington, DC, USA, 2013; pp. 1–25.

education sciences

MDPI

Article

Incorporating Sustainability into Engineering and Chemical Education Using E-Learning

Edmond Sanganyado [1,2,*] and Simbarashe Nkomo [3]

1   Marine Biology Institute, Shantou University, Shantou 515063, China
2   Department of Applied Chemistry, National University of Science and Technology, Ascot, Bulawayo, Zimbabwe
3   Division of Natural Science and Mathematics, Oxford College of Emory University, Oxford, GA 30054, USA; simbarashe.nkomo@emory.edu
*   Correspondence: esang001@ucr.edu; Tel.: +86-150-1727-0075

Received: 15 February 2018; Accepted: 12 March 2018; Published: 23 March 2018

Abstract: The purpose of this study was to develop e-learning activities that could facilitate the integration of sustainability concepts and practices in engineering and chemical education. Using an online learning management system (LMS), undergraduate students in an applied chemistry program at a public university in Zimbabwe participated in an online discussion on the role of chemical reaction engineering in achieving environmental sustainability goals. In the second activity, the students were instructed to prepare a design report for a cost-effective and innovative wastewater treatment plant for a rural hospital. The design report was evaluated through peer review online. Quantitative and qualitative analyses were performed on the two online activities to evaluate student engagement, quality of responses and the incorporation of sustainability into their learning. In the online discussion, 97 comments were made averaging 120 words per comment. Furthermore, the students averaged 3.88 comments, with the majority of comments exhibiting simple and complex argumentation, a deep reflection and widespread use of terms associated with sustainability such as recycling, pollution, waste and the environment. Furthermore, the evaluation of peer reviews revealed that participants demonstrated they could identify the strengths and shortcomings in the design reports. Therefore, this study demonstrated that e-learning, particularly peer review and online discussion, could help chemistry and engineering students appreciate the need for chemical and engineering activities that encourage sustainable development.

Keywords: sustainability; Green Engineering; curriculum development; chemical education; engineering education

1. Introduction

The discharge of chemicals in to the environment and an excessive consumption of natural resources contributed to several global challenges such as climate change, loss of biodiversity, pollution, health risks and pollution [1–7]. In fact, the chemical industry is responsible for the discharge of 98% of CO_2 into the atmosphere, consumption of 78 % of energy, and production of 80,000 different chemicals per year. Thus, there is need for the chemical industry to incorporate sustainability from product development to marketing as well as the end of its life cycle. Sustainability helps to decrease natural resource depletion and chemical discharge in the environment, while increasing the product's economic and social benefits [8]. Engineers and industrial chemists are involved in the discovery, design, development, distribution, and disposal of products. Since they are the primary problem-solvers in the chemical industry, engineers and industrial chemists should have a demonstrable competency in sustainable development [6]. They play a crucial role in finding sustainable solutions of a chemical process at molecular, product, unit operation and plant level by analyzing its environmental, economic and societal impact. Educators play a critical role in ensuring engineers and industrial chemists acquire the essential knowledge, values and basics pertaining to sustainable development. Therefore, sustainability concepts and practices should be introduced into the engineering and chemical education curriculum to equip future chemical industry leaders.

Several engineering societies—including the American Institute of Chemical Engineers, Australia Engineers, Engineers Canada and the Institution of Chemical Engineers—now consider sustainability concepts and practices key engineering competencies (Table 1) [9]. For example, one of the guidelines from Engineers Canada states that engineers "should seek and disseminate innovations that achieve a balance between environmental, social and economic factors while contributing to healthy surroundings in the built and natural environment" [10]. However, engineers can become more skillful in addressing environmental, economic and societal problems in a sustainable way through education. For that reason, in 2005, the United Nations (UN) established the Decade of Education for Sustainable Development aimed at (1) promoting quality education through teaching and learning sustainable development and (2) helping countries attain the millennium development goals (MDGs) through sustainability education [11]. In 2015, the UN further acknowledged the role of education in sustainability via the sustainable development goals (SDGs), SDG 4 in particular [12]. Therefore, it is expedient for higher education practitioners to impart sustainability skills, knowledge and values to future engineers and industrial chemists through the integration of sustainability in engineering education [13,14].

Table 1. Engineering Competencies associated with sustainability.

Engineering Bodies	Competencies
Engineers Canada	C. Conduct engineering activities with an awareness of the associated risk and impact to protect the society, economy and the environment. • Keep all legislation, regulations, codes and standards associated with sustainability. • Identify and assess the negative and positive impacts of all engineering activities. • Identify hazards through evaluation of all safety concerns and the risks of the engineering activities in order to address or mitigate them. • Report all the safety concerns and mitigation strategies to relevant decision-makers.
UK Engineering Council	E3 Undertake engineering activities in a way that contributes to sustainable development. • Operate and act responsibly, taking account of the need to progress environmental, social and economic outcomes simultaneously. • Use imagination, creativity and innovation to provide products and services which maintain and enhance the quality of the environment and community and meet financial objectives. • Understand and secure stakeholder involvement in sustainable development. • Use resources efficiently and effectively.
United States Department of Labor	**Tier 4.** Meet the needs of the present without compromising the ability of future generations to meet their own needs. • Emphasize reducing waste and resource usage while improving efficiency. • Integrate profitability, environmental stewardship and social responsibility. • Ensure industrial processes are designed to reduce adverse environmental impacts. • Leverage technological advances to improve efficiency without compromising the environment.
Engineering Australia	**PE2.2** Understand the social, cultural, global and environmental responsibilities of engineers and the need to employ principles of sustainable development • Understand the interactions and relationships between engineering activities and the social, cultural, environmental, economic and political context they operate. • Appreciate, develop and maintain safe and sustainable systems. • Perform multidisciplinary interactions to broaden knowledge, attain cross-disciplinary goals and maximize integration of engineering activity in the whole project. • Understand economic, societal and environmental risk.

2. Theoretical Background

Several innovative solutions have been used to embed sustainable development in engineering and chemical education. The methods ranged from university-, teacher-, curriculum- and student-oriented approaches. In university-oriented approaches, the leadership created sustainability policies that would be integrated into all activities conducted at the university [15]. For example, the University of Johannesburg employed an innovative management strategy that involved regular planning, policy formulation, brainstorming and benchmarking sessions focused on implementing sustainability goals across the campus [16]. Although ideal, such an approach is difficult to translate to student activities in the classroom without proper teacher training and resources, hence the teacher-oriented approach. There are four approaches that have been identified concerning integrated sustainability in teacher training and these are: (1) diffusing sustainability concepts throughout the curriculum, courses and campus; (2) introduction of dedicated compulsory sustainability subjects; (3) introduction of dedicated sustainability elective subjects; and (4) including a sustainable development component in a core/compulsory subject [17]. The same approaches used in incorporating sustainability into teacher training are often used in curriculum-oriented approaches [18]. In addition, educators have incorporated sustainability into undergraduate laboratory classes by developing experiments and incorporating practices that promote sustainability such as the analysis of bioethanol [19]. using reagents extracted from plants [20]. and implementation of an environmental management system [21]. However, development of a new degree program, course (or class module) or laboratory experiments might be time consuming and costly, particularly in low- and middle-income countries. For that reason, student-oriented approaches, which focus on what the student does, offer an easier and less expensive way for embedding sustainability in engineering education.

Inquiry-based learning methods are the most widely used technique for embedding sustainability. They encourage active learning, as the student takes ownership of their learning thereby promoting the development of higher level thinking skills imperative in sustainable education [22]. A common example of inquiry-based learning employed in engineering education are final year undergraduate research projects. Individual or team research projects are help the students develop and appreciate key engineering competencies [6].Thus, research projects offer a critical bridge between the university and workplace [6,23]. Although integrating sustainability in engineering education is important, there is need for the educator to engage in pedagogical reflection [24]. The objectives of the pedagogical reflection should include establishing the level of understating the learner achieved and the sustainability concepts and practices the learner remembered. Student responses in online activities can be a valuable resource for pedagogical reflection. For example, sustainability encompasses environmental, social or cultural, stakeholders, politics, economic and scientific and multi-disciplinary aspects. Through analysis on the students' responses and grouping the words they used into the six categories would help establish whether the students grasped the interdisciplinary nature of sustainability [24]. An LMS such as Canvas could make it easier to engage in pedagogical reflection as the educator can make formative assessment activities available online.

The increase in internet penetration rate across Africa in the past decade has made e-learning a viable tool for integrating sustainability in higher education [25]. Although e-learning can be challenging in Africa due to student and teacher perceptions and a lack of resources, several studies have found e-learning could improve student engagement and retention as it encourages social interaction and sharing of documents [26,27]. Furthermore, e-learning can be used for harnessing intellectual capital through international collaborative learning [25]. Learning management systems (LMS) used in e-learning such as Canvas, Blackboard and Google Classroom often have features that can be used to measure student engagement. The frequency, duration and regularity with which students use the LMS features can be used to measure student engagement [27]. For example, in a study on cultural and social views on learning, researchers used Canvas, as a platform for measuring social interaction among learners in an online discussion [26]. In an organic chemistry course, peer evaluation of class presentations using Blackboard was shown to help students acquire critical skills in effective communication while encouraging student engagement and interest [28]. Therefore, e-learning can be a useful tool for incorporating sustainability into engineering and chemical education.

Research Question

The study aims to explore the use of e-learning in embedding sustainability concepts and practices in engineering and chemical education. Furthermore, the study aims to develop inquiry-based learning approaches using an online LMS. Hence, the research questions, in context of process engineering, are:

1. How does an online discussion on sustainable development in Zimbabwe foster student engagement?
2. To what extent does online discussion and peer evaluation promote sustainability conscious critical discourse and reflection?

3. Methodology

3.1. Classroom Description

This study was conducted in two process engineering courses over two semesters at a public university in Zimbabwe. The courses comprised of two 120-min lectures each week and three 60-min office hours each week. The enrollment for the courses consisted of final year students studying physical sciences only. The online discussion was conducted over six weeks and the design experiment over two months. Students evaluated the design report of their peers. The students had varying experiences in e-learning. However, none of the students had previously enrolled in a course that used an LMS such as Canvas. Data were collected for 26 students, representing those students who engaged in the discussion and peer evaluation.

3.2. Instructional Strategy

The engineering competencies aligned to sustainability require a high level of knowledge such as designing, creating and integrating [29]. Several studies found that online discussions can encourage critical thinking [30–32]. Therefore, in this study an online discussion was developed that had learning outcomes that addressed the engineering competencies aligned to sustainable development. The discussion question focused on the role of chemical reaction in promoting sustainable development in Zimbabwe. The question was relevant and connected with the students, thus encouraging them to take responsibility for their learning. The second activity required the students to design a wastewater treatment plant for a rural hospital. However, as the future engineers and industrial chemists, the students are expected to be able to evaluate the environment, economic and social impact of a project. Therefore, the students were required to evaluate the reports of their peers by giving a score and writing a corresponding comment justifying the score. Table 2 shows the online activities used in the study and their corresponding learning outcomes. The online discussion posts, design reports and the peer evaluations where submitted on Canvas.

Table 2. Activities used in the incorporation of sustainability into process engineering.

Activities		Learning Outcomes	Source
Type	Question		
Online discussion	According to Milorad P. Dudukovic, "The key challenge for chemical reaction engineering is the development of new more efficient and profitable technologies. This is to be accomplished via an improved science-based scale-up methodology for transfer of molecular discoveries to sustainable nonpolluting processes that can meet the future energy, environmental, food and materials needs of the world." Discuss how chemical reaction engineering can be used to meet the Millennium Development Goals in Zimbabwe, particularly goal 7 on environmental sustainability.	• Apply the fundamentals of reaction engineering in answering sustainability problems. • Develop an understanding of the MDGs and SDGs	Research article [33]

Table 2. *Cont.*

Activities		Learning Outcomes	Source
Type	Question		
Design report	Your local hospital received a report from the Environmental Monitoring Agency that stated that the effluent from the hospital is contaminating a local river. As a design engineer, you are tasked with proposing and designing a wastewater treatment plant for the local hospital. Write a design report for a cost-effective and innovative wastewater treatment plant for the hospital. Discuss how your design is sustainable and helps the nation to meet the MDG 7.	• Apply the fundamentals of reaction engineering in answering sustainability problems. • Evaluate the reliability, effectiveness and limitations of available tools, equipment or technology for solving engineering problems. • Develop a solution that best meets system requirements and specifications.	Report [34]

MDG 7 was used in the activities as it offers a concise reference to the SDGs relevant to process engineering; namely, SDG 6, 7, 9, 13, 14 and 15.

3.3. Classroom Intervention

In a previous study, it was found that students preferred open-ended, rather than debate or case-based, discussions [30]. However, the level of critical thinking was found to be lower in open-ended discussion. Therefore, to encourage critical discourse and reflection in the online discussion, the author gave examples from developed nations on the application of sustainable development to the chemical industry once a week during lectures. Students were encouraged to reflect on how the sustainability practices from developed nations could be tailored for low- and middle-income countries.

3.4. Data Collection

Data was collected after the deadline of the activities had lapsed. The source of the data for this study was student responses from Canvas. The observations made in this study were primarily at the individual level and the whole classroom level. Furthermore, student responses were imported to a word processing software. In the online discussion, student engagement was measured using the length and frequency of their contribution. Since the study was part of normal university activities, no additional ethical review was needed from the Institutional Review Board. To protect the privacy of the participants, data was made available for use of this study only.

3.5. Data Analysis

Data analysis was comprised of determining the level of student engagement in online discussion, evaluating the quality of students' responses and assessing the students' conceptual understanding of sustainability.

3.5.1. Student Engagement

A good learning activity should hold the attention of the learners while encouraging them to participate. Student engagement refers to the psychological investment, time and effort a student puts toward learning. Several studies found that improving student engagement often resulted in an increase in student retention, performance and motivation [35–37]. Furthermore, social interactions with peers, learning content and instructors foster student engagement [38]. Hence, determining the level of student engagement with an online activity can provide evidence on the cognitive development of the students [30]. In this study, behavioral engagement was estimated using the frequency of student participation in the online discussion. The students were instructed to post at least two comments and no upper limit was given. A preliminary estimate of the students' cognitive engagement was estimated using the length of their responses on the online discussion.

3.5.2. Quality of Response

It is important to estimate the quality of the students' argumentation and reflectivity when seeking to establish the effectiveness of a learning approach. However, assessing critical thinking is challenging since it is not a specific ability but a complex set of broad and specific intertwined factors [30]. Critical thinking in online discussions can be assed using the Practical Inquiry Model, which focuses on metacognitive processes rather than the specific learning outcomes [30,39]. In the Practical Inquiry Model, student responses are categorized as triggering, exploration, integration or resolution use different indicators [39]. In this study, a variation of the Practical Inquiry Model was used. Briefly, participant posts from the online discussion (N = 97) were collected from Canvas and subsequently ranked for their quality in five key aspects; namely, Argumentation, Responsiveness, Elicitation, Reflection on Individual Process and Reflection on Group Process (Table 3) [26,32]. The coding was conducted independently by two analysts and the inter-coder reliability was determined.

Table 3. Ranking scheme for quality of participant contribution.

Dimensions	Key Aspects	Rank			
		0	1	2	3
Content	Argumentation	None	Unsupported	Simple	Complex
Discursiveness	Responsiveness	None	Acknowledge	Respond to single idea	Respond to multiple ideas
	Elicitation	None	Unclear question	Question one person	Question whole group
Reflectivity	Reflection in the question or submission	None	Shallow: reflection on own posts with no explanation	Deep: learning process shape one's idea	
	Reflection on group discussion	None	Shallow: reflection on group posts with no explanation	Deep: learning process was shaped by group's idea	

Adapted from Chen et al., 2018.

After submitting the design report, the students anonymously evaluated and scored other participants' reports. Each submission was peer reviewed by three participants. Students were expected to offer constructive criticism of the design report. The quality of the peer evaluation was ranked by establishing if the reviewer identified at least one positive aspect and at least one negative aspect the design report.

4. Results

4.1. Student Engagement

The online discussion was available on Canvas between 21 January 2017 and 28 February 2017. All the participants took part in the activity, generating 97 comments. The students were required to make at least 2 contributions to the discussion. The participants averaged 120 words per comment and each participant averaged 3.88 posts in the online discussion (Figure 1). Of the 26 participants, 8 were highly active, contributing at least 5 comments each. However, 10 participants were less active as they contributed fewer than 3 comments. Eight students made 3 or 4 comments. The participants shared 11 high quality references, comprising reports from universities, government agencies and international organizations and academic papers (data not shown).

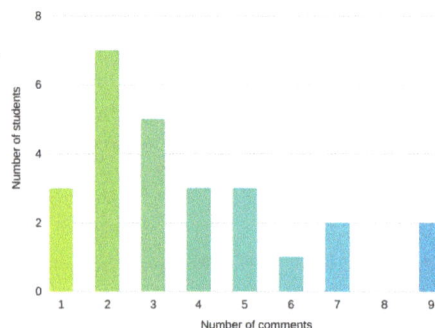

Figure 1. The level of participant engagement in an online discussion.

4.2. Quality of Response

4.2.1. Online Discussion

The differences between the means obtained by the independent coders ranged from 0 to 7.4 %. This suggests the coding scheme had a high repeatability. The quality of the responses and of the argumentation and responsiveness in the comments were slightly above moderate at 2.25 and 2.16, respectively (Table 4). However, the level of elicitation was low at 0.39. They were only 15 questions out of the 97 comments. The reflection in the post and group ideas was slightly above average at 1.24 and 1.14, respectively.

Table 4. The total count and means of weighted conceptual engagement in online discussion.

Key Aspects	Weight				Mean
	0	1	2	3	
Argumentation	6	8	39	44	2.25
Responsiveness	6	16	31	44	2.16
Elicitation	79	4	8	6	0.39
Reflection in the question or submission	16	42	39		1.24
Reflection on group discussion	21	41	35		1.14

4.2.2. Design Report

Out of 26 participants, 22 submitted their design report and were double-blind reviewed by three peers. About 51.5% of the comments were poor, with participants failing to identify the positive or negative aspects of the paper they reviewed. However, 24.2% peer reviews identified at least one positive attribute of a submission. Furthermore, another 24.2% peer reviews identified at least one error in the submission reviewed. The participants identified the errors using expressions such as "lacked," "did not," "neglected," "more research" and "did not include."

4.3. Collaborative Learning

To determine if the participants incorporated sustainability concepts and practices into the online discussion, a word cloud was generated from the forum posts (Figure 2). All the posts from the online discussion were imported to Microsoft Word with a ProWriterAid add-on (Oxford, UK). Redundant words such as the author's last name and the last name of the author of the reading assignment were excluded from the word cloud. The whole class focused primarily on the environmental, economic and scientific aspects of sustainability as words such as environment, waste and chemical were more commonly used. However, the commonly used words did not directly relate to politics, culture and stakeholders.

Figure 2. Visualization of the most frequent words used in the online discussion.

To better understand how the students learned about sustainability using collaborative learning, the questions posed by the participants in the online discussion were analyzed. The participants contributed 15 questions that demonstrated the participants reflected on sustainability. Examples of the engaging questions posted on the forum are shown in Table 5.

Table 5. Elicitations in online discussions.

Examples of Engaging Questions *
1. How does your final year project incorporate the Millennial Development Goal 7?
2. Are there any solvents currently used in Zimbabwe that are persistent in the environment?
3. How exactly does chemical reaction engineering play a role in sustaining a modern Zimbabwean lifestyle?
4. How does chemical reaction engineering affect the environment?
5. Are you implying that most industrial operations in Zimbabwe are using the wrong catalysts? if so, do you have examples to justify that?
6. Considering the research paper was published in 2009, is it possible that basing claims on this text may fail to accommodate some recent developments in reaction engineering?

* Questions have been edited for clarity.

5. Discussion and Conclusions

The study sought to establish e-learning could be used as a tool for incorporating sustainability into engineering and chemical education in developing nations. The results suggested that e-learning activities such as online discussion and peer evaluation of design assignments promoted student engagement. Through analysis of the student comments, it was found that the students were environmental conscious and incorporated principles and practices of sustainability when formulating their arguments or assessing their peers. Such outcomes addressed the key sustainability competencies set by engineering societies such as Engineering Australia, Engineers Canada and Engineering Council.

A high student participation in the open-ended online discussion was observed as the students contributed an average of 3.88 posts per students and 120 words per post. The observed online behavioral patterns suggests the students invested significant time and effort in the activity. Furthermore, 60% of the participants made at least three comments when the required number of posts was two. The Canvas platform fostered social interaction as it possessed social media tools such as responding to other student's comments [26,40]. Furthermore, as students read each other's comments, they were motivated to high-level knowledge processing. However, although the contribution of the whole class was high, 40% of the students were passive submitting the minimum required number of posts or less. Low student participation is often ascribed to the student's digital citizenship. Students who have a low digital proficiency tend to contribute less often. Furthermore, in developing nations lack of internet access or poor connectivity can reduce student participation in online activities. However, in this study the students had free internet access on campus. The author monitored the online discussion without making any contribution. By taking up the role of a facilitator, the instructor can improve student engagement through asking additional questions, clarifying the original questions and addressing any concerns that might arise [31].

The student activities in this study were relevant, authentic and connected with the students as they focused on the societal, economic and environmental challenges in Zimbabwe. Hence, the students were probably motivated to take responsibility of their learning as indicated by the quality of their argumentation and the depth of their reflections. The students demonstrated critical thinking skills because 75% of their comments in the online discussion were built on at least one idea. However, about

40% of the comments showed deep reflection with the students explaining how they were learning sustainable development. However, it was previously observed that when a student considered the online activity to be highly important they became cognitively and emotionally engaged to the task [37]. Cognitive and emotional engagement can be enhanced when the instructor acts as the facilitator who regularly demonstrate the importance of the activity.

The participants managed to incorporate sustainability concepts and practices in their arguments. The word cloud (Figure 2), demonstrated the students included sustainability aspects when formulating their arguments. For example, the following words were widely used; reaction, process, raw materials, inputs, conditions, pollution, which are associated with engineering competencies of foundations of engineering, design, manufacturing and construction, engineering economics, operations and maintenance and safety, respectively [29]. However, the arguments made by the students overlooked political and cultural aspects of sustainable development.

Thus, online learning offered a platform for students to actively learn about sustainability. This study provides instructors with techniques on how to incorporate sustainability into chemical and engineering education. Furthermore, in this study we used and demonstrated techniques for assessing the quality of student responses.

Author Contributions: E.S. conceived, designed and performed the experiments; E.S. and S.N. organized and analyzed the data; E.S. wrote the paper with editorial support from S.N.

Conflicts of Interest: The authors declare no conflict of interest.

References

1. Sanganyado, E.; Lu, Z.; Fu, Q.; Schlenk, D.; Gan, J. Chiral pharmaceuticals: A review on their environmental occurrence and fate processes. *Water Res.* **2017**, *124*, 527–542. [CrossRef] [PubMed]
2. Sanganyado, E.; Rajput, I.R.; Liu, W. Bioaccumulation of organic pollutants in Indo-Pacific humpback dolphin: A review on current knowledge and future prospects. *Environ. Pollut.* **2018**, *237*, 111–125. [CrossRef] [PubMed]
3. Sanganyado, E.; Teta, C.; Masiri, B. Impact of African traditional worldviews on climate change adaptation. *Integr. Environ. Assess. Manag.* **2018**, *14*, 189–193. [CrossRef] [PubMed]
4. Gwenzi, W.; Chaukura, N. Organic contaminants in African aquatic systems: Current knowledge, health risks and future research directions. *Sci. Total Environ.* **2018**, *619–620*, 1493–1514. [CrossRef]
5. Bondarczuk, K.; Markowicz, A.; Piotrowska-Seget, Z. The urgent need for risk assessment on the antibiotic resistance spread via sewage sludge land application. *Environ. Int.* **2016**, *87*, 49–55. [CrossRef] [PubMed]
6. Belu, R.; Chiou, R.; Tseng, T.-L. (Bill); Cioca, L. Advancing sustainable engineering practice through education and undergraduate research projects. In *Education and Globalization*; ASME: Montreal, QC, Canada, 2014; Volume 5, pp. 1–8.
7. Teta, C.; Hikwa, T. Heavy Metal Contamination of Ground Water from an Unlined Landfill in Bulawayo, Zimbabwe. *J. Heal. Pollut.* **2017**, *7*, 18–27. [CrossRef]
8. Allen, D.T.; Shonnard, D.R.; Huang, Y.; Schuster, D. Green Engineering Education in Chemical Engineering Curricula: A Quarter Century of Progress and Prospects for Future Transformations. *ACS Sustain. Chem. Eng.* **2016**, *4*, 5850–5854. [CrossRef]
9. Sheehan, M.; Schneider, P.; Desha, C. Implementing a systematic process for rapidly embedding sustainability within chemical engineering education: a case study of James Cook University, Australia. *Chem. Educ. Res. Pract.* **2012**, *13*, 112–119. [CrossRef]
10. Engineers Canada. *National Guideline on Sustainable Development and Environmental Stewardship for Professional Engineers*; Engineers Canada: Ottawa, ON, Canada, 2016.
11. UNESCO. *United Nations Decade of Education for Sustainable Development (2005–2014): International Implementation Scheme*; UNESCO: Paris, France, 2005.
12. Annan-Diab, F.; Molinari, C. Interdisciplinarity: Practical approach to advancing education for sustainability and for the Sustainable Development Goals. *Int. J. Manag. Educ.* **2017**, *15*, 73–83. [CrossRef]
13. Wright, T.S.A. Definitions and frameworks for environmental sustainability in higher education. *High. Educ. Policy* **2002**, *15*, 105–120. [CrossRef]

14. Hawkins, N.C.; Patterson, R.W.; Mogge, J.; Yosie, T.F. Building a sustainability road map for engineering education. *ACS Sustain. Chem. Eng.* **2014**, *2*, 340–343. [CrossRef]

15. Louw, W. Green curriculum: Sustainable learning at a higher education institution. *Int. Rev. Res. Open Distance Learn.* **2013**, *14*. [CrossRef]

16. Barnard, Z.; Van der Merwe, D. Innovative management for organizational sustainability in higher education. *Int. J. Sustain. High. Educ.* **2016**, *17*, 208–227. [CrossRef]

17. Evans, N. (Snowy); Stevenson, R.B.; Lasen, M.; Ferreira, J.-A.; Davis, J. Approaches to embedding sustainability in teacher education: A synthesis of the literature. *Teach. Teach. Educ.* **2017**, *63*, 405–417. [CrossRef]

18. Verbitskaya, L.A.; Nosova, N.B.; Rodina, L.L. Sustainable development in higher education in Russia: The case of St. Petersburg State University. *High. Educ. Policy* **2002**, *15*, 177–185. [CrossRef]

19. Galgano, P.D.; Loffredo, C.; Sato, B.M.; Reichardt, C.; El Seoud, O.A. Introducing education for sustainable development in the undergraduate laboratory: Quantitative analysis of bioethanol fuel and its blends with gasoline by using solvatochromic dyes. *Chem. Educ. Res. Pract.* **2012**, *13*, 147–153. [CrossRef]

20. Kradtap Hartwell, S. Exploring the potential for using inexpensive natural reagents extracted from plants to teach chemical analysis. *Chem. Educ. Res. Pract.* **2012**, *13*, 135–146. [CrossRef]

21. Montañés, M.T.; Palomares, A.E.; Sánchez-Tovar, R. Integrating sustainable development in chemical engineering education: The application of an environmental management system. *Chem. Educ. Res. Pract.* **2012**, *13*, 128–134. [CrossRef]

22. Pretorius, R.; Lombard, A.; Khotoo, A. Adding value to education for sustainability in Africa with inquiry-based approaches in open and distance learning. *Int. J. Sustain. High. Educ.* **2016**, *17*, 167–187. [CrossRef]

23. Fenner, R.A.; Ainger, C.M.; Cruickshank, H.J.; Guthrie, P.M. Embedding sustainable development at Cambridge University Engineering Department. *Int. J. Sustain. High. Educ.* **2005**, *6*, 229–241. [CrossRef]

24. Lourdel, N.; Gondran, N.; Laforest, V.; Brodhag, C. Introduction of sustainable development in engineers' curricula. *Int. J. Sustain. High. Educ.* **2005**, *6*, 254–264. [CrossRef]

25. Protsiv, M.; Rosales-Klintz, S.; Bwanga, F.; Zwarenstein, M.; Atkins, S. Blended learning across universities in a South-North-South collaboration: A case study. *Heal. Res. Policy Syst.* **2016**, *14*, 67. [CrossRef] [PubMed]

26. Chen, B.; Chang, Y.-H.; Ouyang, F.; Zhou, W. Fostering student engagement in online discussion through social learning analytics. *Internet High. Educ.* **2018**, *37*, 21–30. [CrossRef]

27. Mathews, A.L.; LaTronica-Herb, A. Using Blackboard to Increase Student Learning and Assessment Outcomes in a Congressional Simulation. *J. Polit. Sci. Educ.* **2013**, *9*, 168–183. [CrossRef]

28. Kadnikova, E.N. "Molecules-in-Medicine": Peer-evaluated presentations in a fast-paced organic chemistry course for medical students. *J. Chem. Educ.* **2013**, *90*, 883–888. [CrossRef]

29. United States Department of Labor. *Engineering Competency Model*; United States Department of Labor: Washington, DC, USA, 2015.

30. Richardson, J.C.; Ice, P. Investigating students' level of critical thinking across instructional strategies in online discussions. *Internet High. Educ.* **2010**, *13*, 52–59. [CrossRef]

31. Ouyang, F.; Scharber, C. The influences of an experienced instructor's discussion design and facilitation on an online learning community development: A social network analysis study. *Internet High. Educ.* **2017**, *35*, 34–47. [CrossRef]

32. Wise, A.; Zhao, Y.; Hausknecht, S. Learning Analytics for Online Discussions: Embedded and Extracted Approaches. *J. Learn. Anal.* **2014**, *1*, 48–71. [CrossRef]

33. Dudukovic, M.P. Reaction engineering: Status and future challenges. *Chem. Eng. Sci.* **2010**, *65*, 3–11. [CrossRef]

34. United Nations Economic and Social Council. *Progress towards the Sustainable Development Goals: Report of the Secretary-General*; United Nations Economic and Social Council: New York, NY, USA, 2016.

35. Kahu, E.R. Framing student engagement in higher education. *Stud. High. Educ.* **2013**, *38*, 758–773. [CrossRef]

36. Coates, H. The value of student engagement for higher education quality assurance. *Qual. High. Educ.* **2005**, *11*, 25–36. [CrossRef]

37. Manwaring, K.C.; Larsen, R.; Graham, C.R.; Henrie, C.R.; Halverson, L.R. Investigating student engagement in blended learning settings using experience sampling and structural equation modeling. *Internet High. Educ.* **2017**, *35*, 21–33. [CrossRef]

38. Czerkawski, B.C.; Lyman, E.W. An instructional design framework for fostering student engagement in online learning environments. *TechTrends* **2016**, *60*, 532–539. [CrossRef]

39. Garrison, D.R.; Anderson, T.; Archer, W. Critical inquiry in a text-based environment: computer conferencing in higher education. *Internet High. Educ.* **1999**, *2*, 87–105. [CrossRef]

40. Joksimović, S.; Dowell, N.; Poquet, O.; Kovanović, V.; Gašević, D.; Dawson, S.; Graesser, A.C. Exploring development of social capital in a CMOOC through language and discourse. *Internet High. Educ.* **2018**, *36*, 54–64. [CrossRef]

education sciences

MDPI

Article

Processing Image to Geographical Information Systems (PI2GIS)—A Learning Tool for QGIS

Rui Correia [1], Lia Duarte [1,2,*], Ana Cláudia Teodoro [1,2] and António Monteiro [3]

[1]   Department of Geosciences, Environment and Land Planning, Faculty of Sciences, University of Porto, 4169-007 Porto, Portugal; rui_correia11@hotmail.com (R.C.); amteodor@fc.up.pt (A.C.T.)
[2]   Earth Sciences Institute (ICT), Faculty of Sciences, University of Porto, 4169-007 Porto, Portugal
[3]   Research Center in Biodiversity and Genetic Resources, University of Porto, 4169-007 Porto, Portugal; amonteiro@fc.up.pt
*   Correspondence: liaduarte@fc.up.pt

Received: 26 April 2018; Accepted: 31 May 2018; Published: 6 June 2018

Abstract: Education, together with science and technology, is the main driver of the progress and transformations of a country. The use of new technologies of learning can be applied to the classroom. Computer learning supports meaningful and long-term learning. Therefore, in the era of digital society and environmental issues, a relevant role is provided by open source software and free data that promote universality of knowledge. Earth observation (EO) data and remote sensing technologies are increasingly used to address the sustainable development goals. An important step for a full exploitation of this technology is to guarantee open software supporting a more universal use. The development of image processing plugins, which are able to be incorporated in Geographical Information System (GIS) software, is one of the strategies used on that front. The necessity of an intuitive and simple application, which allows the students to learn remote sensing, leads us to develop a GIS open source tool, which is integrated in an open source GIS software (QGIS), in order to automatically process and classify remote sensing images from a set of satellite input data. The application was tested in Vila Nova de Gaia municipality (Porto, Portugal) and Aveiro district (Portugal) considering Landsat 8 Operational Land Imager (OLI) data.

Keywords: GIS; learning tool; open source software; satellite data

1. Introduction

Inclusive and equitable quality education, and lifelong learning opportunities for citizens are major sustainable development goals to face planetary changes [1]. Education, together with science and technology, is the main driver of progress and transformations of a country [2]. In the engineer area, the professionals can play an important role in industry or in teaching classes as supervisors or mentors. The use of new technologies of learning can be applied to the classroom. This type of interaction increases the motivation in the classrooms [3]. Consequently, computer learning supports meaningful and long-term learning [3]. Therefore, in the era of digital society and environmental issues, a relevant role is provided by open source software and free data that promote universality of knowledge [4]. In this context, Earth Observation (EO) data, remote sensing and Geographical Information Systems (GIS) technologies are crucial to monitor and support the management of natural environment [5]. However, their use to track progress towards global and national targets is still restricted by limited access to open learning and technology [6].

The implementation of open source software in teaching classes allows students to apply different methods and different experimental approaches and share them with other users. This contribute to scientific progress is only possible with the access to unlimited free and open source software [7].

In the education domain, both proprietary and open source GIS software are equally important. Open source software has recently become a stronger player in the GIS field, since it can be accessed, used or modified by their users or developers [8]. Moreover, as proprietary software, it allows exploring the individual GIS functions, such as database management, web mapping, remote sensing data and spatial analysis. However, the license costs, the software technical support and the sustainability are the main differences between open source and proprietary software. In the purpose of GIS/remote sensing classes, both are relevant. The proprietary solutions ensure a professional technical support; and the open source solutions are free, so the students can install and use them in their personal computers, and also have the possibility to modify and distribute the application. The development of System for Automated Geo-Scientific Analyses (SAGA GIS), Geographic Resources Analysis Support System (GRASS GIS) and QGIS are great examples [9–11]. In particular, QGIS software is a significant contributor to the use of EO data and on-line support, e.g., tutorials, forums and code platforms, because of its easy and intuitive use [12].

The development of open source image processing plugins and their inclusion in new capabilities/functionalities in a GIS environment has been the main strategy to tackle inclusiveness in the use of EO data. Plugin development is increasing, and several studies presented different applications/tools with specific objectives in multidisciplinary areas [13–22]. For instance, Garcia-Haro et al. [23] present a teaching tool, which has been designed to enhance the learning in remote sensing. This software is implemented in Interactive Data Language (IDL) language and it is composed of a modular Graphic User Interface (GUI). This tool has been designed in support of remote sensing teaching activities in the University of Valencia. The GUI allows the students to generate several synthetic images that imitate complex ecosystems with several types of trees and shrubs.

EO data are usually used in several teaching areas, such as geology, climate, biology and environment [24,25]. Some open source remote sensing applications are available for that purpose: (i) SAGA GIS software contains a rich library grid, imagery and terrain processing modules [26]; (ii) Orfeo Toolbox (OTB) has a complete image processing library for high spatial resolution data, such as radiometry, Principal Component Analysis (PCA), change detection, pan-sharpening, image segmentation, classification and filtering [27]; (iii) GRASS GIS software contains functions to classification, PCA, edge detection, radiometric corrections, 3 Dimensions (3D) geostatistics analysis and filtering options [28]; (iv) *pktools* are a suite of utilities written in C++ for image processing with a focus on remote sensing applications, relying on the GDAL and OGR [29]; and (v) the Semi-Automatic Classification Plugin (SCP) composed of several functions for optical images processing, which is a powerful package for QGIS software [30].

In image pre-processing and classification, the SCP, developed by Luca Congedo for QGIS, is particularly interesting. It allows semi-automatic classification through a set of supervised classification algorithms for remote sensed images, the calculation of vegetation indices, such as the Normalized Difference Vegetation Index (NDVI) and Enhanced Vegetation Index (EVI), and other image operations, such as the automatic conversion to surface reflectance for different sensors and Region Of Interest (ROI) creation. It also processes data from Landsat, Sentinel-2A, Advanced Spaceborne Thermal Emission and Reflection Radiometer (ASTER) and Moderate Resolution Imaging Spectroradiometer (MODIS).

QGIS software provides some of the tools required to manipulate and analyze remote sensed imagery in *Processing Toolbox* external algorithms (GRASS, SAGA or OTB). However, these tools are dispersed in the software, and some of them don't exist as a single tool (Table 1). A drawback is the manual intervention required to run each single tool, which may be a barrier for non-experts. All operations gathered in a unique application could therefore help to overcome this barrier. The inclusion of the possibility of histogram visualization, filter application, assessment of different image correction techniques, unsupervised classification, and computation of several environmental indices can be a relevant contribution to image processing functionalities in QGIS software. Compared to the SCP plugin, Processing Image To Geographical Information System (PI2GIS) has the advantage of combining a set of image processing procedures in a unique tool. Moreover, the steps to process a satellite image are sequentially implemented/presented in PI2GIS.

Table 1. Image processing functionalities available in QGIS.

Processing Image Method	Tool	Algorithm
Histograms	Not available	Properties of a file (without details)
Filters	OTB	Despeckle (frost, gammamap, kuan, lee), DimensionalityReduction (independent component analysis (ica), maximum autocorrelation factor (maf), noise adjusted principal component analysis (nacpa), pca), Exact Large-Scale Mean-Shift segmentation, step 1 (smoothing), Smoothing (anidif, gaussian, mean)
	SAGA	DTM filter (slope-based), Gaussian filter, Laplacian filter, Majority filter, Morphological filter, Multi direction lee filter, Rank filter, Resampling filter, Simple filter, User defined filter
	GRASS	r.fill.dir, r.mfilter, r.mfilter.fp, r.resamp.filter
DN conversion to reflectance and atmospheric correction	Semi-Automatic Classification Plugin	DOS1
Environmental indexes (NDVI, EVI, NDWI)	SAGA GRASS	Vegetation index (slope-based)—NDVI Enhanced vegetation index—EVI i.vi—NDVI and EVI
Colour composite	GDAL	Merge
Pan-sharpening	OTB GRASS	Pansharpening (bayes, local mean and variance matching (lmvm), Simple RCS Pan sharpening operation (rcs)) i.pansharpen
Unsupervised classification	OTB SAGA GRASS	Unsupervised KMeans image classification K-means clustering for grids i.cluster

The main objective of this work was to contribute to the use of EO data in GIS environment in university teaching courses by creating the PI2GIS, including a new set of image processing operations. This application was implemented in QGIS software using the Python programming language, and it can be divided into three groups of operations: pre-processing, processing, and classification. This application is composed of: (i) creation and histogram visualization; (ii) low, median and high filter application; (iii) atmospheric corrections; (iv) contrast and brightness corrections, including histogram equalization; (v) unsupervised classification algorithm, specifically K-means algorithm; and (vi) the computation of Normalized Difference Water Index (NDWI) in addition to NDVI and EVI. Two Landsat 8 OLI images from Vila Nova de Gaia (VNG) and Aveiro were used to test the application.

2. Methodology

2.1. The PI2GIS Application

PI2GIS was created in the open source QGIS software (version 2.18), licensed by GNU General Public License (GPL), developed in Python language (Python 2.7) and could be installed easily using *Plugins* menu presented in QGIS software [31,32]. QGIS was chosen to develop the application, since it has an Application Programming Interface (API) and Python libraries, such as QGIS API, PyQt4 API and GDAL/OGR library, which support plugin development [33–35]. In order to extract valuable information from EO satellite data, pre-processing and classification steps are essential. Under this circumstance, several operations should be employed: (i) color composite, which consists of a band combination, and it can be false color composite or true color composite; (ii) image enhancement, consisting of the contrast or brightness adjustment, so that the image can be efficiently displayed; (iii) spatial filtering to emphasize some image features; and (iv) image classification in order to group the pixels in land cover classes. The application interacts with the user through three main modules: pre-processing, processing, and classification accessed through a graphic interface and presented individually hereafter. Figure 1 shows the PI2GIS plugin workflow considering all operations involved in the three module groups. The GUI was created with Qt Designer, which is a tool that allows the design and build of GUI (Figure 2) [36].

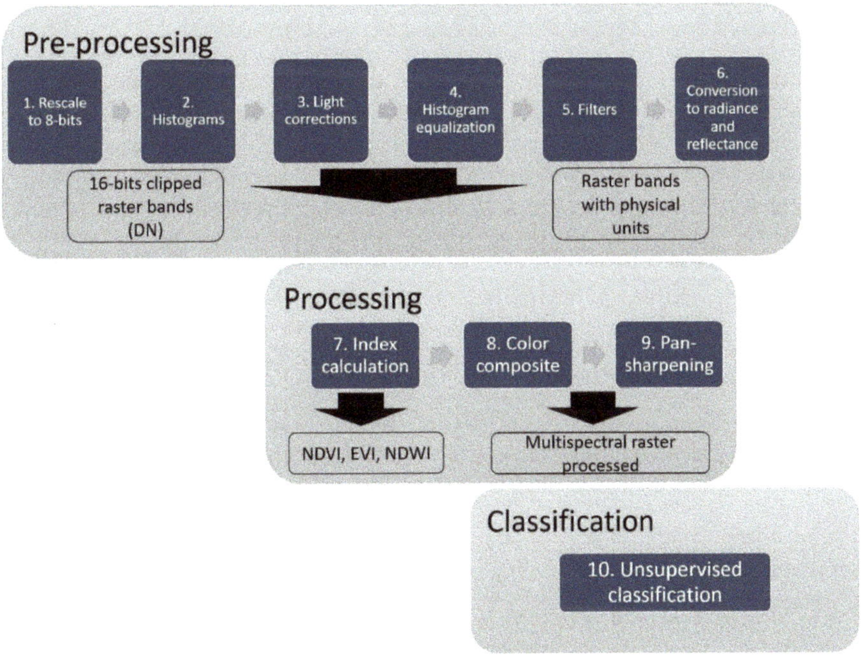

Figure 1. PI2GIS workflow.

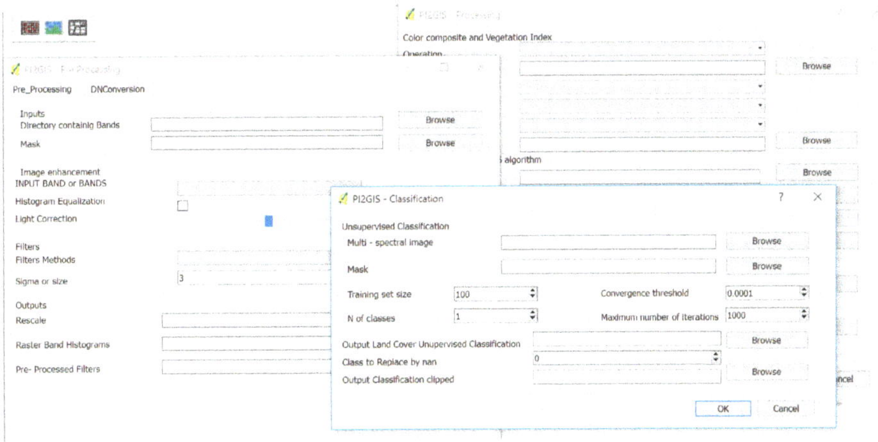

Figure 2. PI2GIS graphic interface.

2.1.1. PI2GIS Pre-Processing Module

The *Pre-processing* module allows data rescaling, image enhancement, radiometric conversion and atmospheric correction. The graphic interface was composed of two tabs: *Pre-processing* and *conversionDN*. The *Pre-processing* tab allows users to deal with image noise by analyzing histograms and enabling the application of histogram equalization or filters. Band histograms were implemented through the function *hist* and created a histogram per image band stored in 8-bits format. To implement this functionality, the Matplotlib, a plotting library for Python was used [37].

In order to reduce the image noise, a group of three operations are available: *Light correction*, *Histogram equalization* and *Filter methods*. These operations can be applied to a single band or to all bands (selected through a combo box). The *Light correction* option allows applying brightness corrections with the support of the widget *QSlider* [34]. By moving the slider, it is possible to increase or decrease the brightness, considering the range between -200 and $+200$. The results are automatically visualized in the QGIS interface through *Preview* button. The *Histogram equalization* option allows improving the image contrast (in 8-bits) and it is implemented with the support of *QcheckBox* widget [34]. This method flattens the grey level histogram of an image in order to equalize all intensities with the aim to normalize image intensity. This transform function is, in this case, a Cumulative Distribution Function (CDF) of the pixel in the image (normalizes the range of pixel values to the desire range). The CDF was applied using the *cumsum()* function from *Numpy* library [38]. To normalize the data, Equation (1) was applied:

$$(255 \times CDF)/CDF[-1] \tag{1}$$

Different types of filters (median, low pass and high pass) were implemented. A median filter is a non-linear filter that allows smoothing the images. This filter is based on a moving-window (matrix), typically with 3×3 size, which moves along the image and recalculates the pixel values by taking the pixels median value. In this application, the *median_filter* function, from *Scipy* library (version 0.19), was used considering a moving-window of 3×3 by default [38]. A low pass filter is a linear filter that smooths the image by removing high frequencies and keeping the low frequencies. *Gaussian_filter*, also from *Scipy* library, with a sigma value of 3, was considered by default (standard deviation for Gaussian Kernel) [38]. A high pass filter was also implemented for contour detection, and it consisted of subtracting the original values of pixels by the values obtained in the low pass filter. It was also based on *Gaussian_filter*, but, in the end, the result was obtained from the subtraction of the original image by the *Gaussian filter* result.

The *conversionDN* tab was intended to perform a conversion from Digital Numbers (DN) to radiance or reflectance considering or not the atmospheric correction (16-bits). In *conversionDN*, a *for* loop defined in a specific directory was used. The bands 1 to 8 of Landsat 8 OLI image are required as input. This function was inspired on a code obtained from GitHub, which uses GDAL and *Numpy* libraries, and all the rescaling factors were based on Landsat 8 Data Users Handbook [39–42]. To perform the DN to Top-Of-Atmosphere (TOA) reflectance conversion, the function calls and reads the metadata file (MTL) available in text format, and uses the *line.split* function (from Python library [32]) to create a dictionary, which links the Landsat 8 band number and the value in the MTL variable for each specific band.

The effects of the atmosphere must be considered to obtain surface reflectance. Therefore, due to the easy computation, the Dark-Object-Subtraction-1 (DOS1) atmospheric correction algorithm was chosen and implemented to improve the estimation of land surface reflectance [43]. The DOS1 is an image-based atmospheric correction defined by Chavez (1996) as follows: "basic assumption is that within the image some pixels are in complete shadow and their radiances received at the satellite are due to atmospheric scattering (path radiance)" [43]. The code implemented was based on the SCP manual [30].

2.1.2. PI2GIS Processing Module

The *Processing* group includes color composite, vegetation index calculation and multi-spectral resolution improvement (pan-sharpening) operations. The *Processing* button allows performing a color composite scene followed by pan-sharpening in order to improve spatial resolution. To perform the pan-sharpening, two tools from OTB were implemented: *Superimpose* sensor and *Pan-sharpening Ratio Component Substitution* (RCS) [27]. The pan-sharpening is a process of merging the panchromatic band with the multispectral bands to create a single high-resolution color image [27]. Different algorithms can be applied. The spatial resolution of input data is one of the major factors that influence the algorithm choice [44]. For instance, Landsat 8 can be combined with Sentinel 2 data for long-term high-frequency monitoring [45]. The *Superimpose* performs the projection of an image into the geometry of another one with the same extension [27]. The *Processing* module also allows the calculation of different environmental indices focusing vegetation and water components: NDVI, EVI and NDWI [46–48]. In the graphic interface, a combo box was created with four options: *color composite, NDVI, NDWI* and *EVI*. In order to create the color composite, the *gdalogr:merge*, algorithm from GDAL library was used [35]. The vegetation indices (Equations (2)–(4)) were implemented through *gdalogr:rastercalculator*, which was an algorithm from GDAL library:

$$NDVI = (NIR - RED)/(NIR + RED) \tag{2}$$

$$EVI = G \times (NIR - RED)/(NIR + C1 \times RED - C2 \times BLUE + L) \tag{3}$$

$$NDWI = (GREEN - NIR)/(GREEN + NIR) \tag{4}$$

where NIR, RED, BLUE and GREEN are the near-infrared reflectance surface, red reflectance surface, blue reflectance surface and green reflectance surface, respectively; L (L = 1) is a canopy background adjustment term; C1 (C1 = 6) and C2 (C2 = 7.5) are the coefficients of the aerosol resistance term and G is a gain or scale factor (G = 2.5) [46].

2.1.3. PI2GIS Classification Module

The *Classification* group was defined with the aim to perform an unsupervised classification with the color composition image obtained from the *Processing* group. The *Classification* button incorporates the K-means algorithm available in the OTB library, which has the advantage of not requiring the definition of training classes [27]. K-means is an unsupervised classification algorithm that solves the clustering problem. Clustering is the process of finding a structure in a collection of unlabeled data by organizing objects, of which members are similar in some way. K-means is an algorithm, which slits the image into different clusters of pixels in the feature space, and each of them defined by its center and pixel is allocated to the nearest cluster [49]. In this interface, it is necessary to define a set of parameters which are existent in the OTB algorithm: training set size (clusters size, value 100 by default) is part of the original image that is used to train the model, the convergence threshold (value 0.0001 by default) is assessed when the K-means algorithm has converged on a good solution and should be stopped, the maximum number of iterations (value 1000 by default) and the number of classes. To support the definition of those parameters, the widgets *QspinBox* and *QdoubleSpinBox* were used [34]. The input image is the resulting multispectral high-resolution image from the pan-sharpening process obtained through RCS. As a result of the unsupervised classification, several classes were defined according to their spectral properties. The algorithm considered *Nodata* values as a class. To remove the *Nodata* values, the *gdalwarp* was used [35].

2.2. Demonstration of PI2GIS

Dataset and Study Area

A satellite scene from the Landsat 8 OLI (date: 22 July 2016; path: 204, Row 32), delivered in *tiff* format (16 bits) by the *Earth Explorer—U.S. Geological Survey* [50] and projected on Universal Transverse Mercator (UTM)-WGS84 29N (EPSG: 32,629), was used to demonstrate the applicability and evaluate PI2GIS performance. The scene was re-projected to the official Portuguese coordinate system Portugal Transverse Mercator 2006—European Terrestrial Reference System 1989 (PTTM06—ETRS89, EPSG: 3763). The municipality of VNG (168 km^2) and Aveiro district (2800 km^2) are located in Northwest Portugal (Figure 3). The VNG municipality belongs to Porto district and is mainly occupied by urban, industrial and rural land uses. The elevation ranges from a sea level to 261 m with rugged terrain. Aveiro district is characterized by flat terrain with the majority of the territory above 100 m of altitude and the coast with about 40 km width in the South.

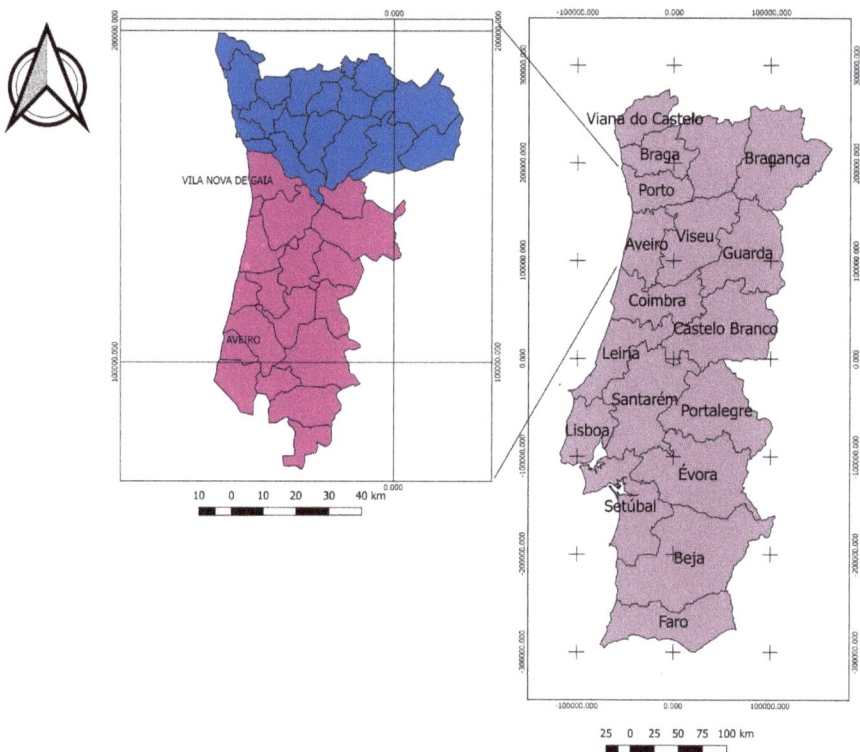

Figure 3. VNG and Aveiro location.

3. Results

3.1. Pre-Processing Module

To test the *Pre-processing* group, the band 2 (B2-blue band) of Landsat 8 OLI image was selected. In this group, the images were rescaled from 16-bits to 8-bits. Then, the histogram for each band was created. The histogram creation was easy, and the user only needed to choose one band or all the bands. The histograms were created and immediately shown, in addition to being saved in a specific chosen folder. Figure 4 presents the B2 histogram for each study area.

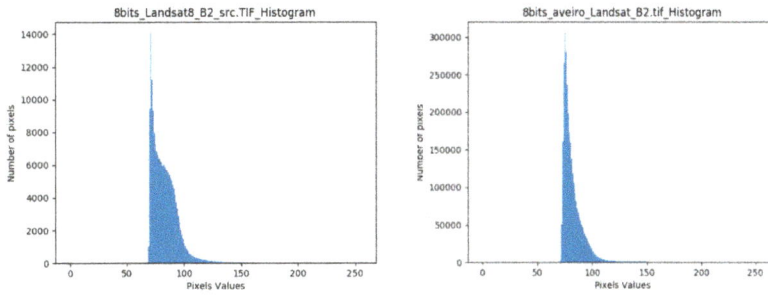

Figure 4. Histogram of Landsat B2 from VNG (**left**) and Aveiro (**right**).

The histogram equalization was also tested, as well as all the filter types. Figure 5 presents the comparison between (a) the original 8-bits image, (b) with the image resulted from the histogram equalization, examples of (c) increasing brightness, (d) a low pass filter, (e) a median pass filter and (f) a high pass filter.

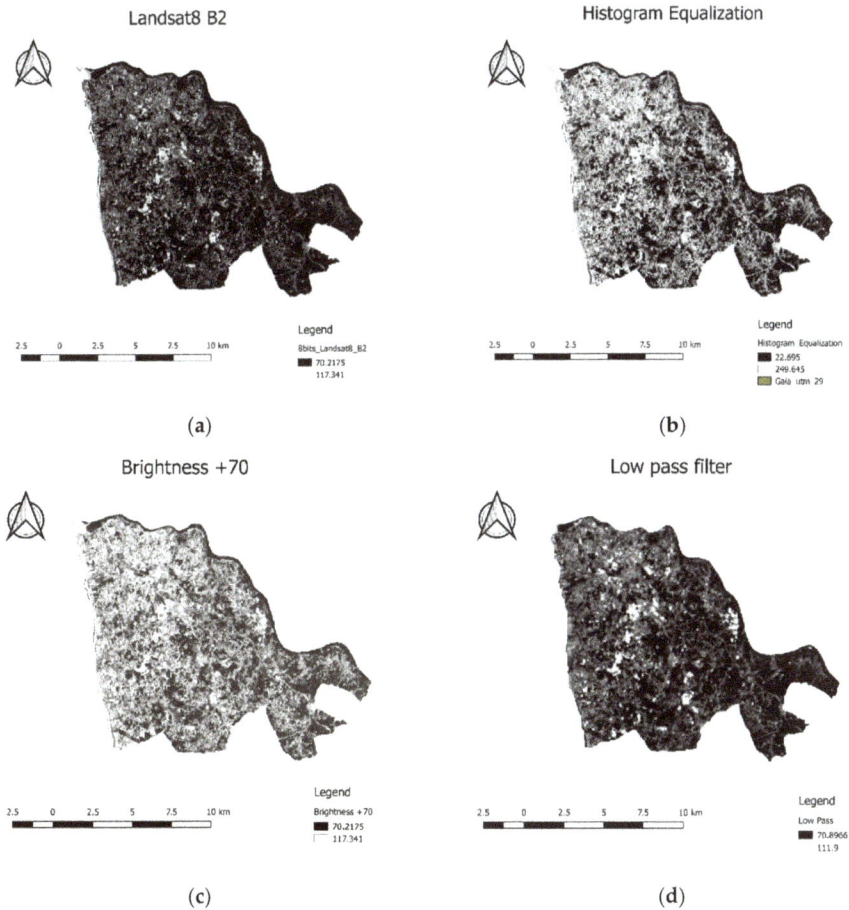

Figure 5. *Cont.*

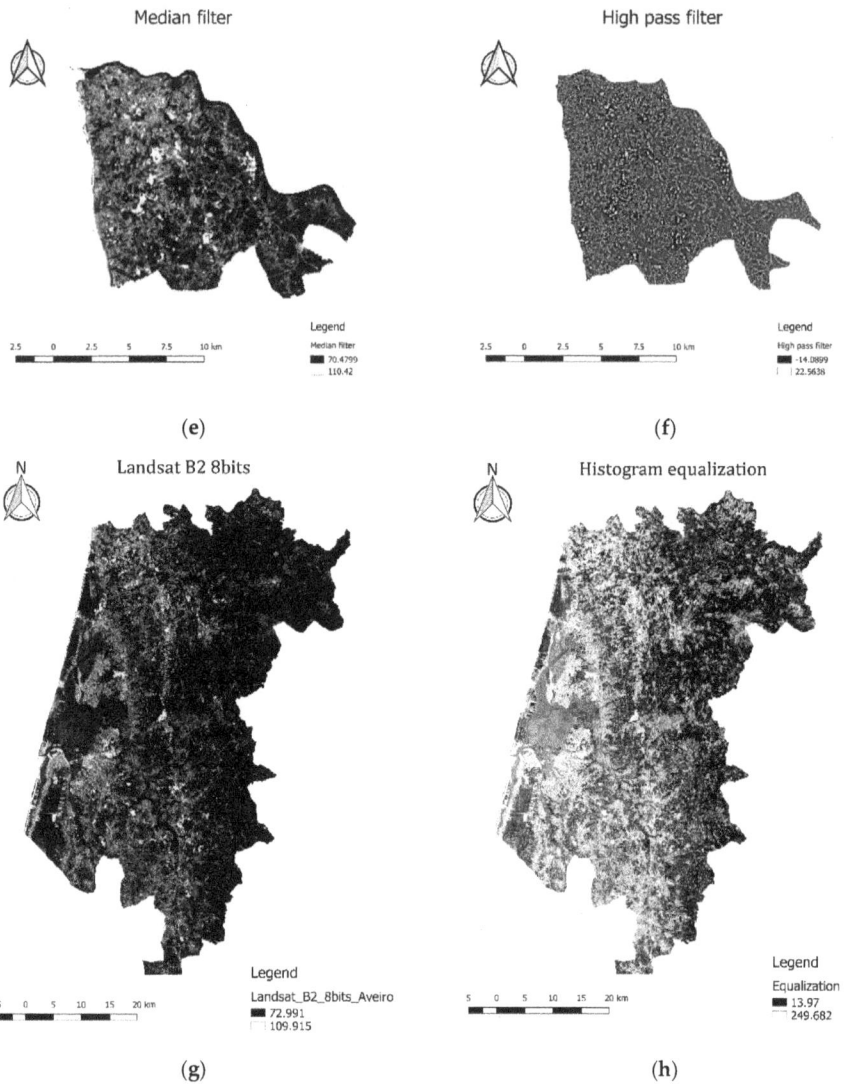

(e)

(f)

(g)

(h)

Figure 5. *Cont.*

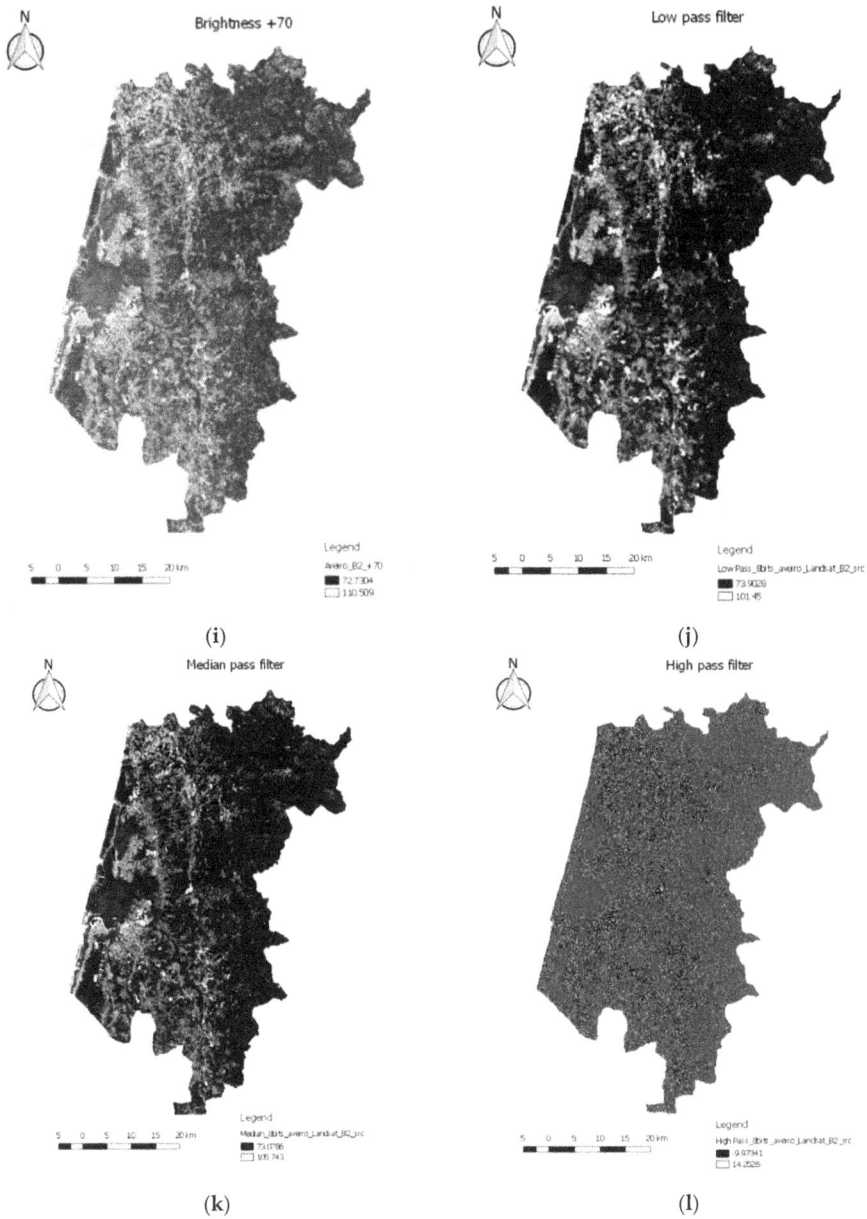

Figure 5. (a) Landsat B2; (b) histogram equalization of Landsat B2; (c) Landsat B2 with increasing brightness; (d) Landsat B2 with a low pass filter applied; (e) Landsat B2 with a medium pass filter applied; (f) Landsat B2 with a high pass filter applied to VNG; and (g–l) the same procedures applied to Aveiro district.

A zoom was applied considering the low pass filter application to VNG. Figure 6 illustrates the differences between the original B2 image and the image after the filter application. As expected, by analyzing Figure 6, it is possible to verify a smoothing effect.

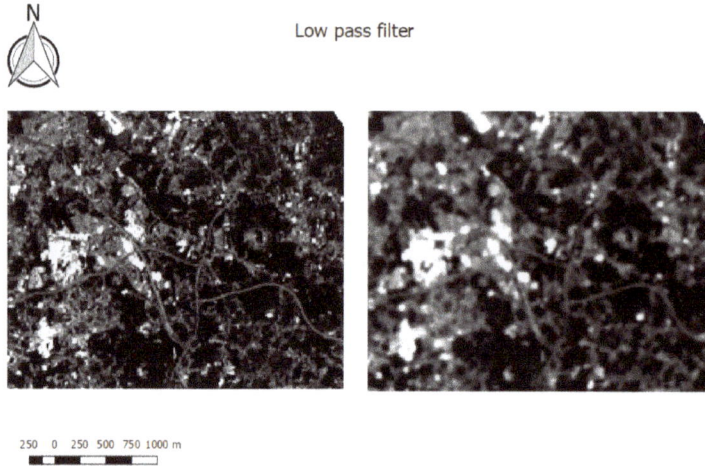

Figure 6. Zoom before- (**left**) and after-the application (**right**) of the low pass filter.

3.2. Processing Module

In the *Processing* group, RGB composite, NDVI, EVI and NDWI were generated (Figure 7). NDVI and EVI presented higher values in the spring and summer season. The image considered was obtained in July 2016. EVI presented lower values compared to NDVI, as EVI corrected some distortions in the reflected light caused by the particles in the air, as well as the ground cover below the vegetation [50]. NDWI was designed to maximize the reflectance of water. Thus, Douro River and several water bodies were easily identified.

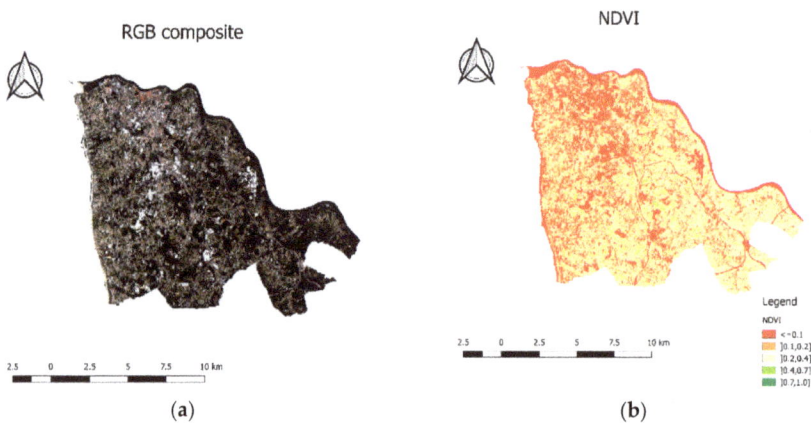

(a)

(b)

Figure 7. *Cont.*

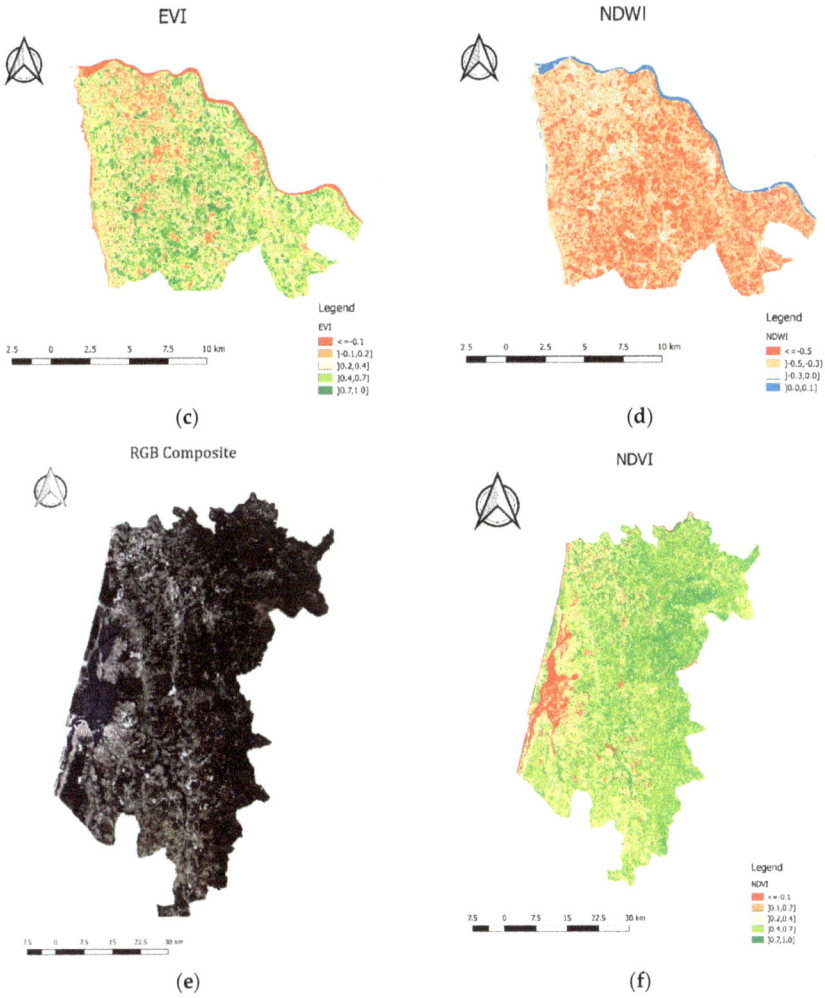

Figure 7. *Cont.*

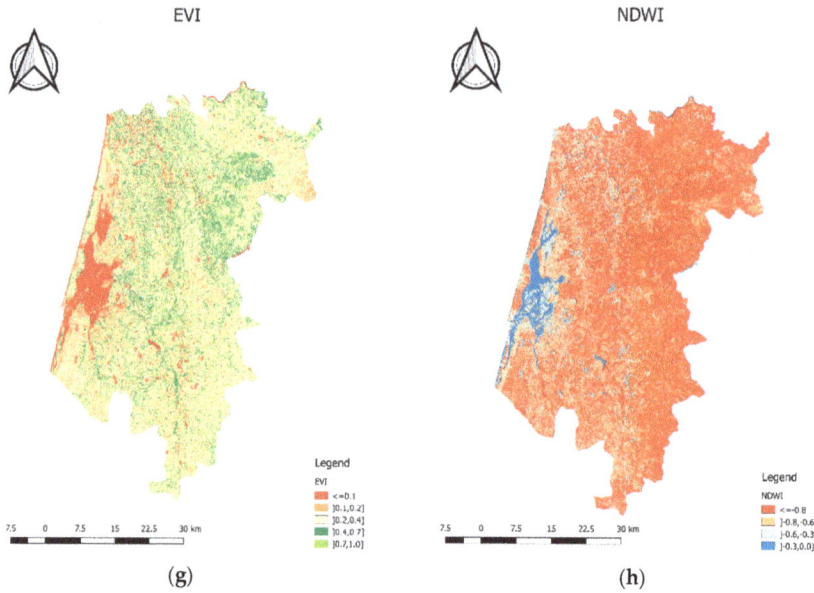

Figure 7. (**a**) RGB composite; (**b**) NDVI map; (**c**) EVI map; (**d**) NDWI map applied to VNG; and (**e–h**) the same procedures applied to Aveiro.

After obtaining the indices, the pan-sharpening functionality was tested. Figure 8 illustrates before and after pan-sharpening operation for VNG. The input files used were the multispectral image obtained using the color composite tool with 30 m of spatial resolution and the panchromatic band with 15 m of spatial resolution. By mere visual inspection, it is possible to identify the differences between the original multispectral lower-resolution image (30 m) and the pan-sharpening (15 m) image obtained by merging the original bands with the high-resolution panchromatic band.

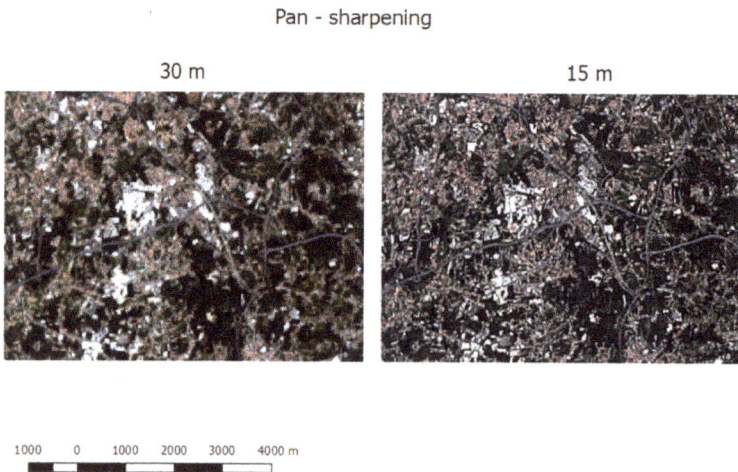

Figure 8. Pan-sharpening example.

3.3. Classification Module

The classification module was tested by performing an unsupervised classification, considering the K-means algorithm and the RGB combination (RGB543) as input (Figure 9). In the false color composition, the vegetation regions are displayed in red (Figure 9). Unsupervised classification is the only classification option available in the current version of PI2GIS (Figure 10).

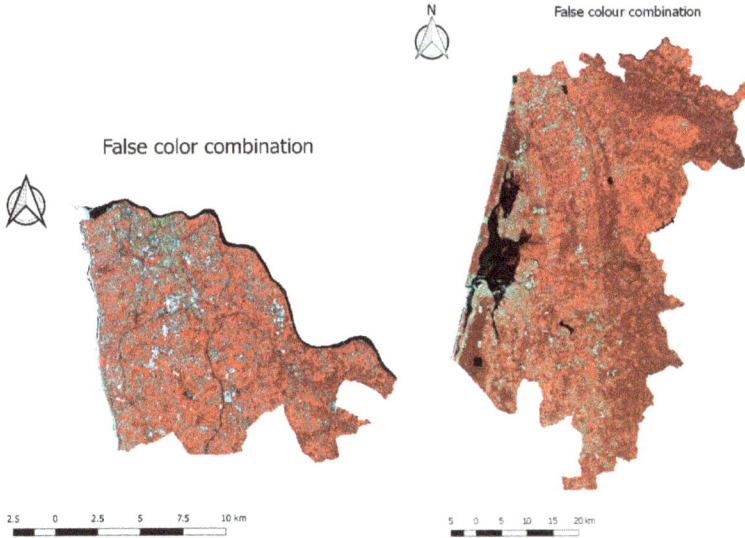

Figure 9. False color combination (RGB543) for VNG (**left**) and Aveiro (**right**).

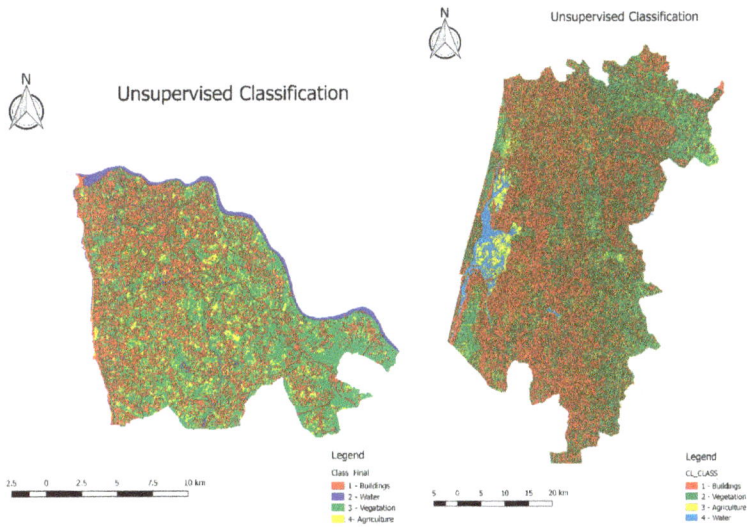

Figure 10. Unsupervised classification for VNG (**left**) and Aveiro (**right**).

4. Discussion and Conclusions

This study proposed PI2GIS, a new plugin, as an integrated and friendly open source tool to process satellite imagery in QGIS, an open source environment. PI2GIS was created as a new tool to support the remote sensing course at an university level, and it can be a useful contribute to the easy and modern education, since it was developed under the open source concept and it is expected to be simpler to use than other remote sensing software, considering the positive feedback given by the master degree students that have already tested it. PI2GIS emerged as a new teaching tool to improve the usage of EO data. The usage of new technologies, such as open source software, can stimulate the students to gain responsibility in their profession of engineers, to work independently with new modern tools, and to apply and adapt their knowledge to unexpected new situations, such as the necessity to create a new application (e.g. PI2GIS). This work provides two approaches in new learning methods: the possibility of development of GIS open source applications and the use of these new tools in the classroom. PI2GIS has been tested with a few Master Degree students in remote sensing, who verified the limitations and drawbacks of the tool, and gave us the feedback. Moreover, the application was already improved. In the future, PI2GIS will be used in the remote sensing Master Degree classes.

PI2GIS was developed to reduce barriers considering high education in environmental sciences and more specifically in the use of GIS and remote sensing data to monitoring environmental change. A free and open source tool for basic remote sensing operations, available everywhere, also in less developed countries, can be a great contribution to equity issues, and will help to prepare young generations for environmental changes.

PI2GIS is available in www.fc.up.pt/pessoas/liaduarte/PI2GIS.rar. The attempt was to contribute to inclusive and equitable education and lifelong learning on environmental management and sustainable development, since this application can be used in remote sensing classes, in which some of the students are not familiar with remote sensing specific software. Therefore, a tool composed of the main processing image algorithms can be very useful to teach the main steps of satellite image processing.

The plugin modules were individually tested and PI2GIS can be considered operational. All expected outputs were created correctly with Landsat-8 satellite imagery data, and VNG municipality and Aveiro district were chosen to test the application. Therefore, the user-friendly interface of PI2GIS is a valid option to open learning and technology to monitoring and management of natural environment. Furthermore, PI2GIS takes advantage of all the GIS and remote sensing algorithms presented in the *Processing Toolbox*. Therefore, in terms of the user profile of PI2GIS, this can be a great advantage. In addition, the easiness in the creation or/and improvement of applications, such as PI2GIS, enhanced its performance. PI2GIS is not only a plugin with a set of QGIS algorithms. PI2GIS includes several procedures that are not so intuitive when QGIS is used.

Certainly, PI2GIS will be improved in the future, adding new functionalities and considering the feedback given by the users. For instance, in the *Pre-processing* group, the data processing, such as the transformation to PTTM06—ETRS89 (EPSG: 3763) or any other coordinate system, could be included. Further enhancement methods to improve contrast could also be included, such as optimal linear transformation or Gama correction. Furthermore, PI2GIS intended to process Sentinel-2 images. The calculation of other indices, such as Soil Moisture Index (SMI) and ISODATA method for unsupervised classification, are also being implemented in PI2GIS.

This work proposed PI2GIS as an open integrated and user-friendly remote sensing application implemented in QGIS software to remote sensing imagery analysis. The integration and releasing of remote sensing imagery operations through a user-friendly interface may constitute an added value of PI2GIS, regarding sparse open solutions available to deal with imagery processing in a GIS environment. The open access of PI2GIS contributes towards the millennium sustainable development goals, including access to equitable quality education and lifelong learning opportunities for citizens. PI2GIS can contribute to the quality of education by simplification of the usage remote sensing data in a GIS environment in environmental management and teaching. The graphic interface is very friendly, especially to non-familiarized users with GIS/remote sensing software, so the basic remote sensing data can be generated and interpreted. PI2GIS aims to contribute and not to invalidate existing GIS/remote sensing tools.

Author Contributions: R.C., A.C.T. and L.D. conceived and designed the experiments; R.C. performed the experiments; A.C.T. and L.D. analyzed the data; A.M. contributed reagents/materials/analysis tools; R.C., A.C.T., L.D. and A.M. wrote the paper.

Conflicts of Interest: The author declares no conflicts of interest.

References

1. United Nations, Economic and Social Council. Progress towards the Sustainable Development Goals Report of the Secretary-General. 2017. Available online: https://unstats.un.org/sdgs/files/report/2017/secretary-general-sdg-report-2017--EN.pdf (accessed on 15 March 2018).

2. Zaldívar-Colado, A.; Alvarado-Vázquez, R.I.; Rubio-Patrón, D.E. Evaluation of Using Mathematics Educational Software for the Learning of First-Year Primary School Students. *Educ. Sci.* **2017**, *7*, 79. [CrossRef]

3. Conradty, C.; Bogner, F.X. Hypertext or Textbook: Effects on Motivation and Gain in Knowledge. *Educ. Sci.* **2016**, *6*, 29. [CrossRef]

4. Khan, A.W. Universal Access to Knowledge as a Global Public Good. Global Economic Symposium, 2009. Available online: https://www.globalpolicy.org/social-and-economic-policy/global-public-goods-1-101/50437-universal-access-to-knowledge-as-a-global-public-good.html (accessed on 15 March 2018).

5. Bush, A.; Sollmann, R.; Wilting, A.; Bohmann, K.; Cole, B.; Balzter, H.; Martius, C.; Zlinszky, A.; Calvignac-Spencer, S.; Cobbold, C.A.; et al. Connecting Earth observation to high-throughput biodiversity data. *Nat. Ecol. Evol.* **2017**, *1*, 0176. [CrossRef] [PubMed]

6. Doldirina, C. Open Data and Earth Observations. The Case of Opening up Access to and Use of Earth Observation Data through the Global Earth Observation System of Systems. *Open Data Earth Obs.* **2015**, *6*, 73. Available online: https://www.jipitec.eu/issues/jipitec-6-1-2015/4174/doldirina.pdf (accessed on 15 March 2018).

7. Rocchini, D.; Petras, V.; Petrasova, A.; Horning, N.; Furtkevicova, L.; Neteler, M.; Leutner, B.; Wegmann, M. Open data and open source for remote sensing training in ecology. *Ecol. Inform.* **2017**, *40*, 57–61. [CrossRef]

8. Tsou, M.-H.; Smith, J. Free and Open Source Software for GIS Education. Department of Geography, San Diego State University, 2011. Available online: http://www.geotechcenter.org/uploads/2/4/8/8/24886299/tsou_free-gis-for-educators-whitepaper-final-draft-jan281.pdf (accessed on 21 November 2017).

9. Di Palma, F.; Amato, F.; Nolè, G.; Martellozzo, F.; Murgante, B. A SMAP Supervised Classification of Landsat Images for Urban Sprawl Evaluation. *ISPRS Int. J. Geo-Inf.* **2016**, *5*, 109. [CrossRef]

10. Huth, J.; Kuenzer, C.; Wehrmann, T.; Gebhardt, S.; Tuan, V.Q.; Dech, S. Land Cover and Land Use Classification with TWOPAC: Towards Automated Processing for Pixel—And Object-Based Image Classification. *Remote Sens.* **2012**, *4*, 2530–2553. [CrossRef]

11. Usha, M.; Anitha, K.; Iyappan, L. Landuse Change Detection through Image Processing and Remote Sensing Approach: A Case Study of Palladam Taluk, Tamil Nadu. *Int. J. Eng. Res. Appl.* **2012**, *2*, 289–294.

12. QGIS Development Team. QGIS Geographic Information System. Open Source Geospatial Foundation Project, 2017. Available online: https://www.qgis.org/en/site/ (accessed on 21 November 2017).

13. Becker, D.; Willmes, C.; Bareth, G.; Weniger, G.-C. A plugin to interface openmodeller from QGIS for species' potential distribution modelling. *ISPRS Ann. Photogramm. Remote Sens. Spat. Inf. Sci.* **2016**, *3*, 251–256. [CrossRef]

14. Jiang, Y.; Sun, M.; Yang, C. A Generic Framework for Using Multi-Dimensional Earth Observation Data in GIS. *Remote Sens.* **2016**, *8*, 382. [CrossRef]

15. Jung, M. LecoS—A python plugin for automated landscape ecology analysis. *Ecol. Inform.* **2016**, *31*, 18–21. [CrossRef]

16. Teodoro, A.C.; Duarte, L. Forest fire risk maps: A GIS open source application—A case study in Norwest of Portugal. *Int. J. Geogr. Inf. Sci.* **2013**, *27*, 699–720. [CrossRef]

17. Duarte, L.; Teodoro, A.C.; Gonçalves, J.A.; Guerner Dias, A.J.; Espinha Marques, J. A dynamic map application for the assessment of groundwater vulnerability to pollution. *Environ. Earth Sci.* **2015**, *74*, 2315–2327. [CrossRef]

18. Duarte, L.; Teodoro, A.C.; Gonçalves, J.A.; Soares, D.; Cunha, M. Assessing soil erosion risk using RUSLE through a GIS open source desktop and web application. *Environ. Monit. Assess.* **2016**, *188*. [CrossRef] [PubMed]

19. Duarte, L.; Teodoro, A.C. An easy, accurate and efficient procedure to create Forest Fire Risk Maps using Modeler (SEXTANTE plugin). *J. For. Res.* **2016**, *27*, 1361–1372. [CrossRef]

20. Duarte, L.; Teodoro, A.C.; Maia, D.; Barbosa, D. Radio Astronomy Demonstrator: Assessment of the Appropriate Sites through a GIS Open Source Application. *ISPRS Int. J. Geo-Inf.* **2016**, *5*, 209. [CrossRef]

21. Duarte, L.; Teodoro, A.C.; Gonçalves, J.A.; Moutinho, O. Open-source GIS application for UAV photogrammetry based on MicMac. *Int. J. Remote Sens.* **2016**, *38*, 8–10. [CrossRef]

22. Duarte, L.; Teodoro, A.C.; Moutinho, O.; Gonçalves, J.A. Distributed Temperature Measurement in a Self-Burning Coal Waste Pile through a GIS Open Source Desktop Application. *ISPRS Int. J. Geo-Inf.* **2017**, *6*, 87. [CrossRef]

23. Garcia-Haro, F.J.; Martinez, B.; Gilabert, M.A. An educational software for remote sensing. In Proceedings of the 10th International Technology, Education and Development Conference (INTED), Valencia, Spain, 7–9 March 2016; pp. 5072–5079.

24. Wang, X.; Chen, N.; Chen, Z.; Yang, X.; Li, J. Earth observation metadata ontology model for spatiotemporal-spectral semantic-enhanced satellite observation discovery: A case study of soil moisture monitoring. *GISci. Remote Sens.* **2016**, *53*, 22–44. [CrossRef]

25. Song, Y.; Wu, C. Examining human heat stress with remote sensing technology. *GISci. Remote Sens.* **2018**, *55*, 19–37. [CrossRef]

26. SAGA GIS. SAGA Software. 2017. Available online: http://www.saga-gis.org/ (accessed on 21 November 2017).

27. OTB (OrfeoToolbox). Pan-Sharpening Description. 2017. Available online: https://www.orfeo-toolbox.org/Applications/Pansharpening.html (accessed on 22 November 2017).

28. Nikolakopoulos, K.; Oikonomidis, D. Quality assessment of ten fusion techniques applied on Worldview-2. *Eur. J. Remote Sens.* **2015**, *48*, 141–167. [CrossRef]

29. Mandanici, E.; Bitelli, G. Preliminary Comparison of Sentinel-2 and Landsat 8 Imagery for a Combined Use. *Remote Sens.* **2016**, *8*, 1014. [CrossRef]

30. GRASS GIS. 2017. Available online: https://grass.osgeo.org/ (accessed on 22 November 2017).

31. Pktools. Pktools Documentation. 2017. Available online: http://pktools.nongnu.org/html/index.html (accessed on 16 March 2018).

32. Congedo, L. Semi-Automatic Classification Plugin Documentation. 2016. Available online: https://fromgistors.blogspot.com/p/semi-automatic-classification-plugin.html (accessed on 21 November 2017). [CrossRef]

33. GNU Operating System. GNU General Public License. 2017. Available online: https://www.gnu.org/licenses/gpl-3.0.en.html (accessed on 21 November 2017).

34. Python. Python Programming Language. 2017. Available online: http://python.org/ (accessed on 22 November 2017).

35. QGIS API. QGIS API Documentation. 2017. Available online: http://www.qgis.org/api/ (accessed on 20 November 2017).

36. PyQt4 API. PyQt Class Reference. 2017. Available online: http://pyqt.sourceforge.net/Docs/PyQt4/classes.html (accessed on 20 November 2017).

37. GDAL. Geospatial Data Abstraction Library. 2017. Available online: http://www.gdal.org/ (accessed on 20 November 2017).

38. Qt Designer. Qt Documentation, Qt Designer Manual. 2017. Available online: http://doc.qt.io/qt-4.8/designer-manual.html (accessed on 21 November 2017).

39. Matplotlib. 2017. Available online: http://matplotlib.org/ (accessed on 22 November 2017).

40. Numpy. 2017. Available online: http://www.numpy.org/ (accessed on 22 November 2017).

41. Scipy. 2017. Available online: https://www.scipy.org/ (accessed on 21 November 2017).

42. Solem, J.E. Programming Computer Vision with Python. Creative Commons, 2012. Available online: http://programmingcomputervision.com/downloads/ProgrammingComputerVision_CCdraft.pdf (accessed on 21 November 2017).

43. Gomez-Dans, J. Landsat DN to Radiance Script Using GDAL and Numpy. 2017. Available online: https://gist.github.com/jgomezdans/5488682 (accessed on 20 November 2017).

44. Landsat 8 Data Users Handbook. USGS, 2017. Available online: https://landsat.usgs.gov/sites/default/files/documents/Landsat8DataUsersHandbook.pdf (accessed on 21 November 2017).

45. Chavez, P.S. Image-Based Atmospheric Corrections—Revisited and Improved Photogrammetric Engineering and Remote Sensing, [Falls Church, Va.]. *Am. Soc. Photogramm.* **1996**, *62*, 1025–1036.

46. Reed, B.C.; Brown, J.F.; VanderZee, D.; Loveland, T.R.; Merchant, J.W.; Ohlen, D.O. Measuring phenological variability from satellite imagery. *J. Veg. Sci.* **1994**, *5*, 703–714. [CrossRef]

47. Huete, A.; Justice, C.; Van Leeuwen, W. MODIS Vegetation Index (MOD 13). *Algorithm Theor. Basis Doc.* **1999**, *3*, 213.

48. Gao, B.-C. NDWI A Normalized Difference Water Index for Remote Sensing of Vegetation Liquid Water from Space. *Remote Sens. Environ.* **1996**, *58*, 257–266.

49. Lillesand, T.; Kiefer, R.W.; Chipman, J. *Remote Sensing and Image Interpretation*, 7th ed.; Wiley: New York, NY, USA, 2015; 768p, ISBN 978-1-118-34328-9.

50. USGS. U.S. Geological Survey. 2017. Available online: https://www.usgs.gov/ (accessed on 21 November 2017).

education sciences

MDPI

Article

A Virtual Resource for Enhancing the Spatial Comprehension of Crystal Lattices

Diego Vergara [1,*], **Manuel Pablo Rubio** [2] **and Miguel Lorenzo** [3]

1   Technological Department, Catholic University of Ávila, 05005 Avila, Spain
2   Construction Department, University of Salamanca, 37008 Salamanca, Spain; mprc@usal.es
3   Department of Mechanical Engineering, University of Salamanca, 37700 Salamanca, Spain; mlorenzo@usal.es
*   Correspondence: diego.vergara@ucavila.es or dvergara@usal.es; Tel.: +34-920-251-020

Received: 31 July 2018; Accepted: 18 September 2018; Published: 21 September 2018

Abstract: Students commonly exhibit serious spatial comprehension difficulties when they come to learning crystal systems. To solve this problem, an active methodology based on the use of a Didactic Virtual Tool (DVT)—developed by the authors—is presented in this paper. The students' opinion was obtained from a survey carried out on 40 mechanical engineering students. The analysis of the obtained results reveals that, by using this DVT, students achieve a better understanding of the contents where spatial difficulties often arise during conventional teaching. Several DVT features were highly valued by the students, e.g., didactic use was rated 9.5 out of 10 and the methodology using the DVT in the classroom was rated 8.5 out of 10. In addition, the results revealed two factors that the students considered essential for using a DVT, both related to the tool design: (i) the modern aspect, i.e., it is necessary to keep a DVT updated to avoid obsolescence; and (ii) the DVT must be appealing in order to attract the students' attention.

Keywords: crystal system; Bravais lattices; spatial abilities; didactic virtual resources; didactic virtual tools; design; active methodology

---

1. Introduction

One of the key contents in most of the subjects related to Materials Science and Engineering is the spatial atoms arrangement in crystal systems [1–5]. Commonly, this topic is of the highest interest for engineering and chemical students. However, diverse shortcomings related to the spatial vision arise: the visualization of a spatial atom arrangement from different points of view; the visualization of the cross sections that identify the atomic arrangement from different points of view; visualization of the positions of the octahedral and tetrahedral interstitial sites of the crystal systems; etc.

Previous papers highlighted how important and necessary it is to acquire a good spatial visualization skill for the future professional life of an engineer or architect [6]. In this sense, serious spatial visualization difficulties in technical or engineering subjects were detected [7–10]. On the other hand, the advantage of using virtual tools (VTs) in the teaching of subjects included in engineering degrees have been demonstrated many times [11–14]. Thus, teaching experiences carried out with engineering students revealed that an improvement of both the acquired knowledge and spatial visualization skill are achieved by using a VT. This is in agreement with previous studies [15–20] where it was demonstrated that the visualization skill can be improved through appropriate training. Taking all the previously mentioned into account, two facts can be stated: (i) the graphical design of any Didactic Virtual Tool (DVT) is a key factor in achieving an educative aim; and (ii) in the case that such a tool is used for solving any spatial visualization problem, the tool design becomes an even more influential factor.

Regarding the topic of this paper—spatial visualization of atom arrangement in crystal systems—several examples of didactic apps using Information and Communication Technologies (ICTs) exist [21–24]. Some of them are available on open-access websites [25]. The DVT presented in this paper is based on a 2D environment, but this DVT includes the features of a 3D interaction in real time with the crystal systems, i.e., students can freely interact with all the crystal systems in order to spatially understand the position of each atom within the lattice. In this way, the potential problems involved in spatial visualization related to this type of teaching are solved. This type of interactivity is really important from the teaching point of view, according to previous studies that revealed a link between this feature and the students' interest in learning [26].

As a summary, this paper presents a teaching methodology based on a DVT which can be applied to the teaching of crystal systems in engineering degrees. Furthermore, the students' opinions are reflected, revealing a link between the DVT design and the motivation generated in the students for continuing to use the DVT.

2. Didactic Virtual Tool (DVT)

The DVT presented in this paper was designed using Unity®. This commercial software allows the development of videogames and 3D interactive applications in real time. Nowadays, many types of software can be used for designing interactive tools in real time, e.g., Quest3D®, OGRE®, GameStudio®, etc. Among them, the authors selected Unity® as it offers the following advantages: (i) Unity® offers content for diverse platforms such as PC, Mac, Nintendo, Wii, and most of all, IOS (Iphone) and Android (in the present day, Android is the most used platform on mobile devices or tablets); (ii) it allows a high variety of script languages for programming; (iii) modular application growth is allowed; (iv) the software documentation is complete; (v) it offers a free licence, this way applications can be developed free of charge from the beginning.

The developed DVT was designed from the teaching flow shown in Figure 1. According to this, the didactic contents included in the DVT appear in a logical sequence in various screens. The main screen helps students to understand and to visualize the differences between the different types of atom arrangements in metals (seven crystal systems). From this screen, students have access to the different subclasses, completing the 14 Bravais Lattices. In each one of them, four visualization modes are available for enhancing the interaction between student and DVT: the unit cell with two visualization options (real and expanded), and the global lattice (several unit cells assembled) also with real and expanded views.

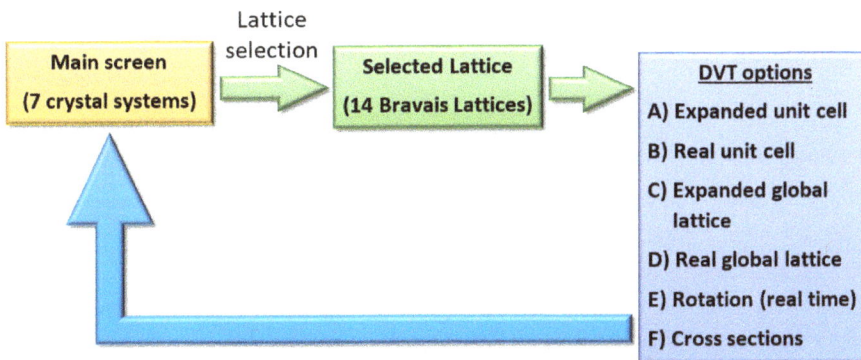

Figure 1. Didactic virtual tool (DVT) teaching flow.

According to the authors' teaching experience, these visualization options are useful for achieving a better understanding of the spatial visualization of the crystal systems. On one hand, the unit cell helps students to understand the placement of atoms individually, and on the other hand, the global lattices are useful for understanding the crystal system as a whole. For each one of 14 Bravais lattices, the expanded view is the first approach that makes the understanding easier, but the real view visualization option is the one that students must understand as a final goal. The different lattice types are ordered in the DVT from the easiest to the more complex with the four visualization options available: (i) expanded unit cell; (ii) real unit cell; (iii) expanded global lattice; and (iv) real global lattice. In addition, for each one of the 14 Bravais lattices, students can rotate and make cross sections of the crystal lattice in order to enhance the spatial visualization. In this way, students can visualize and easily understand the spatial arrangements of atoms in the diverse crystal systems from the simplest to the most complex by themselves.

Figure 2 shows the DVT main screen including the seven crystal systems: (i) Cubic; (ii) Tetragonal; (iii) Orthorhombic; (iv) Rombohedral or Trigonal; (v) Hexagonal; (vi) Monoclinic; and (vii) Triclinic. By clicking on any of them, a new screen appears revealing the respective 14 types of Bravais lattice: (i) Simple Cubic, BCC, and FCC; (ii) Simple Tetragonal and Body Centred Tetragonal; (iii) Simple Orthorhombic, Body-Centred Orthorhombic, Face-Centred Orthorhombic and Base-Centred Orthorhombic; (iv) Rombohedral; (v) Simple Hexagonal; (vi) Simple Monoclinic and Base-Centred Monoclinic; (vii) Triclinic. As an example, Figure 3 shows the three cubic crystal systems which appear after clicking the cubic system button on the main screen (Figure 1): (i) simple cubic; (ii) body centred cubic (BCC); and (iii) face centred cubic (FCC). In this way, students can explore the fourteen possibilities of atom arrangement in 3D (Bravais Lattices) included in the seven different crystal systems (main screen) on their own.

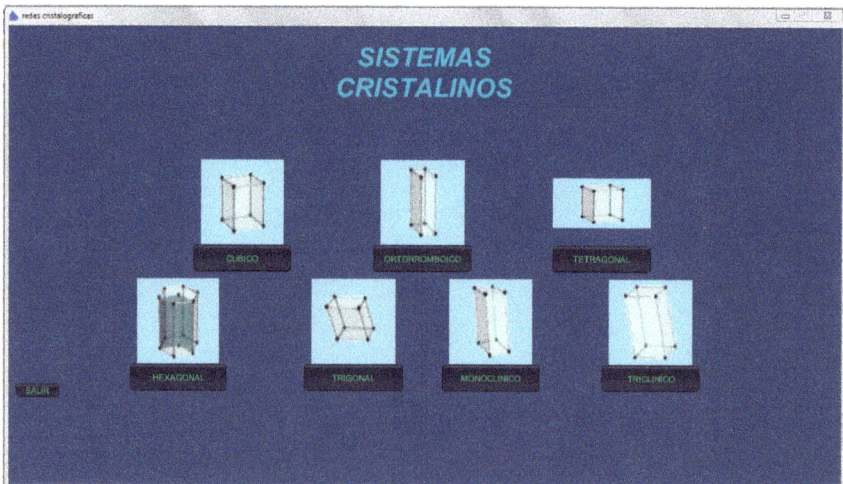

Figure 2. Main screen of the Didactic Virtual Tool showing the seven crystal systems.

The study of crystal lattices is included in the course of all subjects related to Materials Science and Engineering. Thus, the DVT includes the common crystal lattice classifications used in any textbook of Materials Science and Engineering [1–5]. It is necessary to highlight that the common hexagonal close-packed structure (HCP) is not included here since, according to the textbooks, this one cannot be classified within the hexagonal crystal system: "the so-called HCP does not correspond to the hexagonal system, since this one only admits a simple cell. In fact, it is equivalent to a simple rombohedric cell, easier to be represented" (sentence translated from Spanish in Reference [27] (p. 46)).

By using the DVT, users can search for any of these crystal systems and their respective lattices in an easy and intuitive way. In the DVT, the atomic arrangement in crystalline solids is represented placing atoms in the interaction points of a three-dimensional structure with virtual lines (expanded unit cell, Figure 4). In this way, the spatial understanding of the crystal systems in the initial stages is simplified and the teaching becomes easier. Once the atoms arrangement is understood, the DVT serves for visualizing the real atoms position, in which the atoms are in contact with each other (real unit cell, Figure 5a), and even for visualizing the complete crystal lattice formed by several unit cells (real global lattice, Figure 5b). This DVT also includes theoretical information relevant to each unit cell (see the right upper side on Figure 4): geometrical data of the angles and sides of the unit cell, coordination number, atomic packaging factor, and number of atoms per cell.

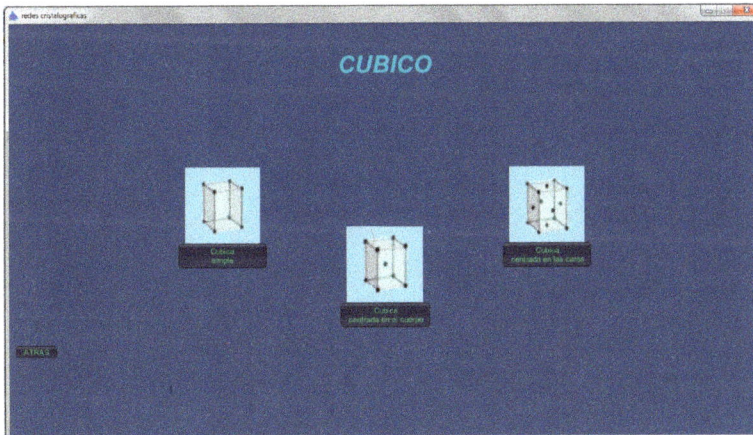

Figure 3. Screen of the Didactic Virtual Tool showing crystal cubic systems.

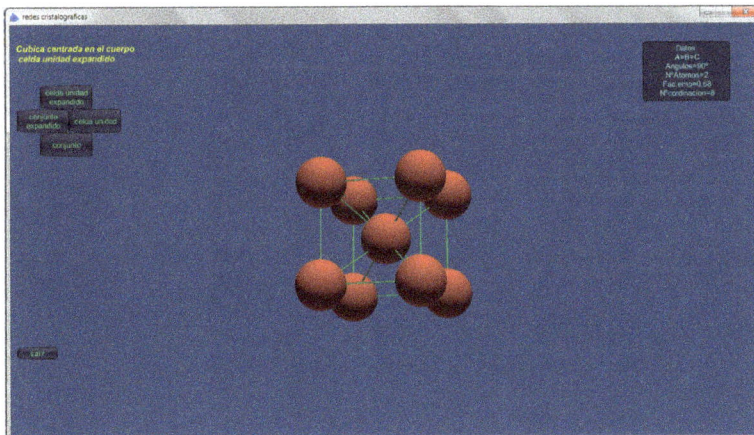

Figure 4. Options included in the DVT for each spatial lattice: expanded unit cell.

For the teaching–learning process, the most relevant issue is that students *interact in real time* with the unit cell or crystal system by using the DVT. The interactivity feature in DVTs is really helpful for the user spatial comprehension [28]. With this in mind, the DVT allows each lattice to be rotated, turned, and placed where the user desires (Figure 6). Furthermore, the DVT allows students to perform cross sections of the lattice to enhance the understanding of the spatial position of the crystal system. As an example, Figure 7 shows diverse cross sections of a spatial lattice corresponding to a BCC (Figure 7a) and FCC (Figure 7b). From these figures the reader can easily comprehend that the spatial visualization of this type of lattice is not an easy task, and thus he/she can also understand the difficulties that can arise when this issue is explained without a DVT.

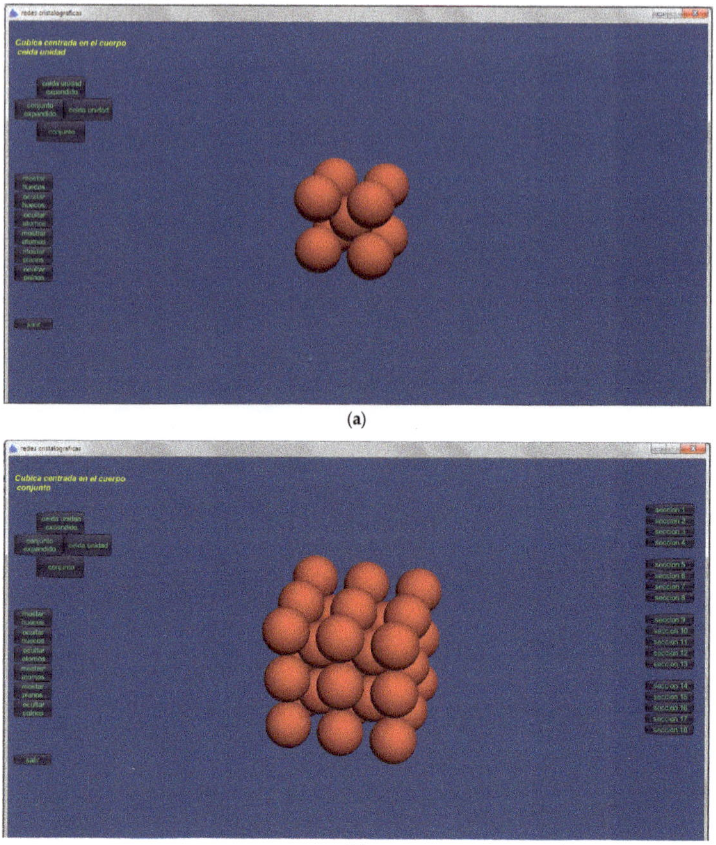

(a)

(b)

Figure 5. Options included in the DVT for each spatial lattice: (**a**) real unit cell; (**b**) real global lattice.

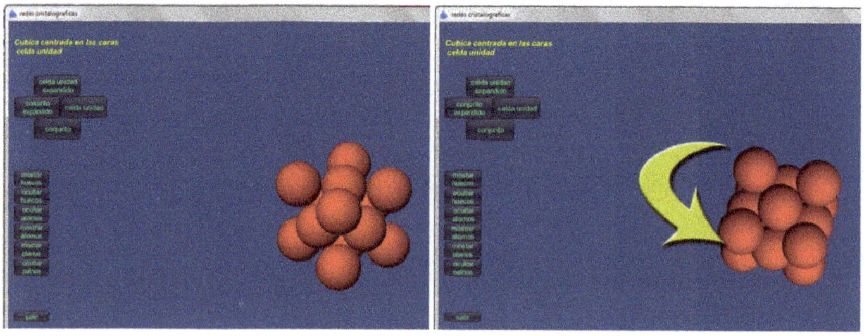

Figure 6. Rotation of a unit cell by using the interactivity feature in real time.

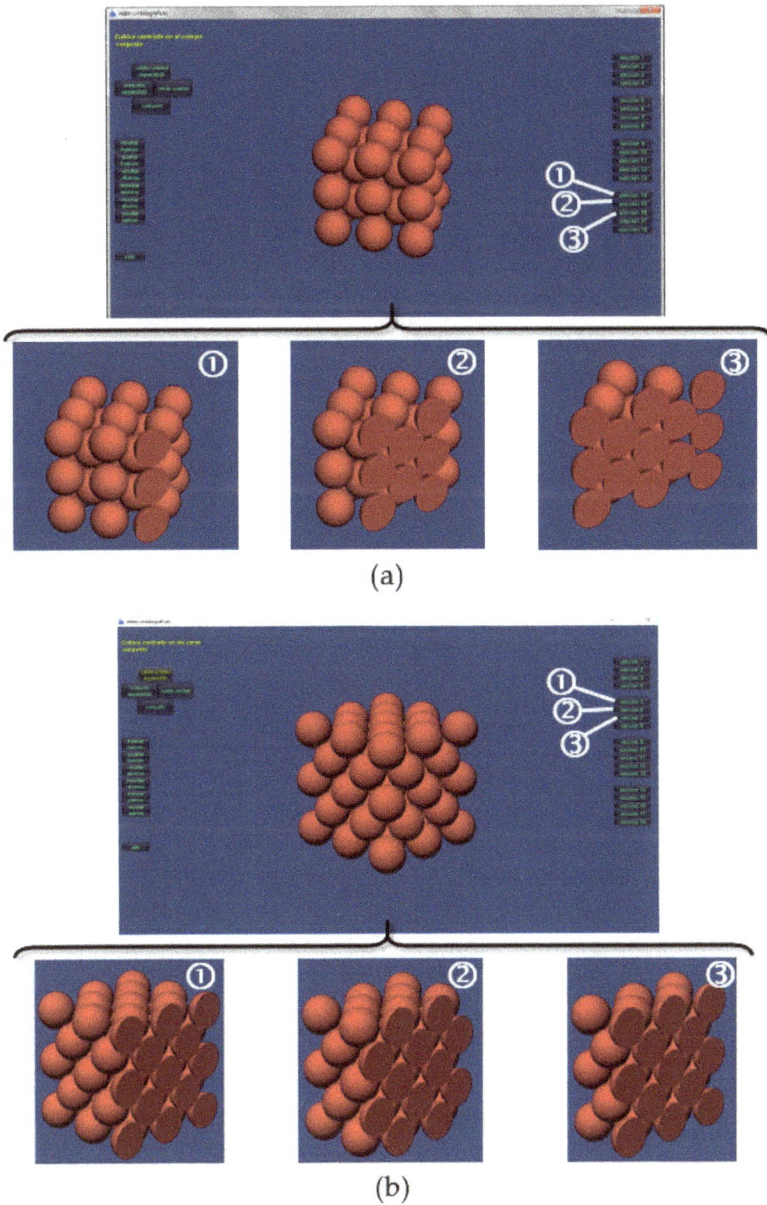

Figure 7. Cubic crystal structures: (**a**) general view and three cross sections of the body centred cubic (BCC) lattice; (**b**) general view and three cross sections of the face centred cubic (FCC) lattice.

3. Methodology

Active learning, as compared with the traditional style based on passive methodologies, involves a higher work effort for both students and instructors. Despite this, active learning experiences are positively valuated by both instructors and students [29,30]. The methodology presented in this paper for the teaching–learning process of crystal systems is based on active learning, following the stages depicted in Table 1: (i) master class; (ii) DVT by means of a self-leaning process; (iii) problem solving by means of cooperative learning.

Table 1. Phases of the proposed methodology.

Phase	Time	Description
1	2 h	**Theoretical explanation of the crystal systems**: explanation of the theory basics of the topic, formulation, crystallographic directions, Miller indexes for crystallographic planes, etc.
2	0.5 h	**Application of the DVT**: spatial understanding of each one of the crystal systems and Bravais lattices. The use of the Didactic Virtual Tool (DVT) will be developed on an individual level to enhance a self-learning process.
3	2 h	**Exercises**: students will solve a collection of exercises in small groups of 3–4 students, enhancing a collaborative learning process and peer-learning.

Firstly, the instructor teaches a master class of approximately 2 h demonstrating all the theoretical basic concepts of the topic. In this phase, the traditional passive teaching must be avoided, aiming for a dynamic response from the students (active methodology). To achieve this goal, the authors have developed some incomplete notes that students must complete as the master class is carried out. In this way, students stop being mere spectators as they must be actively involved in the learning environment.

Following this, students interact with the DVT to spatially understand the crystal systems. Authors consider that the best option is to develop this phase on an individual level, since each student needs a different time to reach a spatial understanding of this type of crystal lattice. To make this self-learning process effective, the DVT must be intuitive and easy to use.

Finally, the instructor will deliver a collection of exercises to be solved in small groups (3–4 students) to enhance the cooperative learning process. This is the last stage of the proposed methodology and it is of paramount importance that the learning process is as efficient as possible, thus the authors consider that working in groups to solve individual doubts is the best option. Equally, according to previous studies, small groups of 3–4 people can contribute to a more efficient cooperative learning [31], thereby enhancing the peer-learning process. According to previous studies [32], this is an aspect that facilitates the resolution of spatial geometric problems. In Figure 8, a scheme of the proposed methodology is represented.

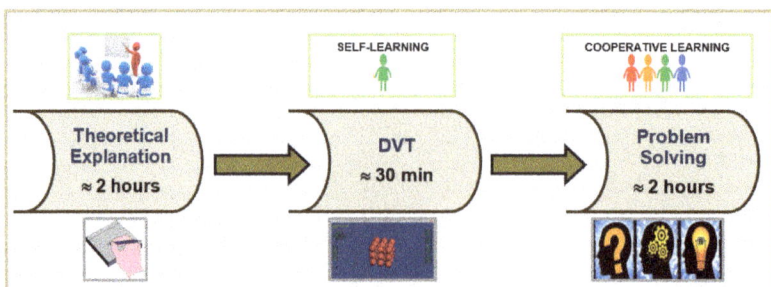

Figure 8. Scheme of the proposed methodology.

4. Students' Opinions

This section includes the students' opinions after using the previously explained DVT. To do so, a quasi-experimental research was carried out on a total of 40 students enrolled on the Materials

Science and Engineering course (second course in the Mechanical Engineering Degree). Some of the questions included in the survey are directly related to the design of the tool and others to the methodology applied (Table 2). The survey results show the importance of the design of DVTs (Figure 9). On one hand, the DVT was well rated with regard to its didactic usefulness, as is reflected by the average values (9.5 points out of 10 with a standard deviation of 0.33), and its ease of use (9 points out of 10 with a standard deviation of 0.28). Thus, the results show that the students consider that the DVT is an intuitive application. On the other hand, the results reveal that the tool does not motivate all the students to the same level (7.5 points out of 10 with a standard deviation of 0.67). In all the answers, the standard deviation is low, but in the case of learning motivation, the deviation is twice that obtained in the other questions. These results suggest that students' motivation is in essence subjective, and in addition, they demonstrate that what is highly motivating for one student may not rouse the same interest in others.

Table 2. Questions included in the students' survey.

Number	Question	Response
1	Rate from 1 to 10 the following DVT features:	(A) Interactivity (B) Ease of use (C) Didactic usefulness (D) Motivation (E) Design
2	Rate from 1 to 10 the methodology proposed for learning crystalline systems.	(A) Master class (B) DVT (C) Problem solving
3	Possible improvements of the DVT	Comments:

Motivation in learning is one of the most analyzed topics in education [33–35] as it seems to be linked with academic performance. For this reason, the results of this survey which are related to this aspect deserve to be analysed in detail. The relatively low motivation can be explained in two possible ways: (i) students do not feel motivated by the topic itself; and (ii) the DVT does not motivate students due to an unattractive design (the rate was 8 points out of 10, Figure 9). Both factors are possible, however, taking into account the following facts: (i) this topic, crystal systems, is included in the final exam of the subject; and (ii) the DVT can help students to spatially understand all the crystal lattices, the authors consider that the most influential factor for the observed low student motivation is the tool design. In addition, considering that students had previously handled other (more sophisticated) DVTs related to virtual labs of Material Science [36–39], the comparison between them and this DVT could be the main cause of the low valuation given to the design feature of the exposed DVT (the lowest one in Figure 9). Therefore, *a direct relationship exists between the design of a VT and the motivation generated in users to keep on using it.* Consequently, both aspects (design and motivation) are really important in DVTs [40,41] and they should be taken into account in the early design phase of any didactic tool.

The valuation given by students of the other DVT features is really promising, given that all of them (interactivity, ease of use, and didactic use) are rated between 9 and 9.5 out of 10 (Figure 9). These results suggest that the DVT presented in this paper—in a global sense—helps students to acquire a better spatial comprehension of crystal systems. Furthermore, the authors verified that the presented DVT helps instructors to teach crystal systems, since the contents involved in such a topic are difficult to explain orally. This way, the DVT accomplishes its intended function.

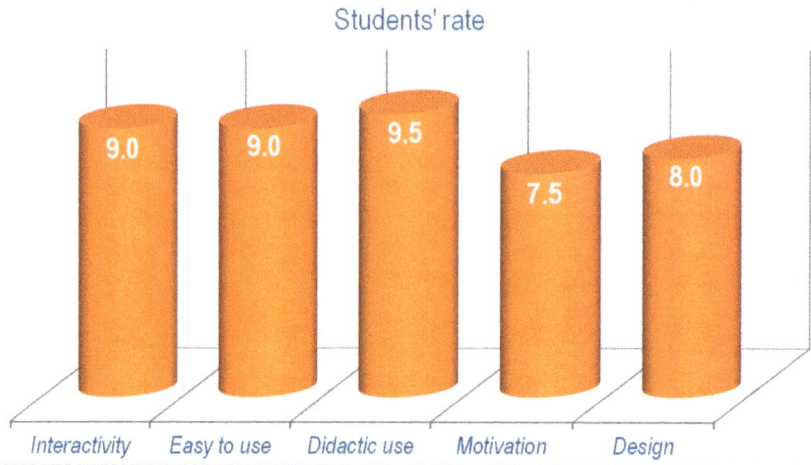

Figure 9. Ratings of students' answers to the survey.

Additionally, the survey results suggest that students are in agreement with the methodology used (Figure 8), given that they have evaluated all the phases—master class, DVT, and problem solving—with a rating higher than 8 out of 10. The phase with the highest rating (slightly higher than the others) is the last one, where students cooperate among themselves. The ratings given for each phase (out of 10) are as follows: 8.5 master class, 8.5 DVT, and 9 problem solving. Finally, with regards to the third question included in Table 2, students also highlight that the following improvements could be made:

- Amend small typos;
- Indicate the place of octahedral and tetrahedral interstitial sites;
- Include the structure HCP in the DVT;
- Implement the application of the Miller indexes and the crystallographic directions.

The authors consider the previously depicted students' improvement suggestions to be very useful and interesting, and hence, are trying to implement all of them in a new version of the DVT. Furthermore, they are updating the aesthetic of the tool with the ultimate aim of enhancing learning motivation in students.

5. Conclusions

A Didactic Virtual Tool (DVT) is presented in this paper. This didactic tool, developed with the software Unity®, can be used in the teaching–learning process of crystal systems. Taking into account the difficulty of spatial understanding related to this topic, the presented DVT helps student to visualize the spatial arrangements of atoms, thereby making the teaching process easier.

From the developed survey, it was confirmed that *a direct relationship exists between the virtual tool design and the motivation generated in the user to keep on using it*. Therefore, in this paper the relevance of the design of any didactic virtual resource to make it not only educative but also motivating is highlighted. To achieve this goal, it is recommended to conduct student surveys periodically to detect any decrease in motivation, and if necessary, to carry out the appropriate improvements in a new version of the DVT using the most modern commercial software possible.

Author Contributions: D.V. and M.P.R. design the virtual tool; D.V., M.P.R. and M.L. analyzed the data and wrote the paper.

Funding: This research was funded by the University of Salamanca, grant number ID2015/0267.

Conflicts of Interest: The authors declare no conflict of interest.

References

1. Asbhy, M.F.; Jones, D.R.H. *Engineering Materials 2: An Introduction to Microstructures and Processing*, 4th ed.; Editorial Butterworth Heinemann: Oxford, UK, 2013.
2. Callister, W.D. *Materials Science and Engineering: An Introduction*, 7th ed.; John Wiley & Sons: New York, NY, USA, 2007; ISBN 978-0-471-73696-7.
3. Mangonon, P.L. *The Principles of Materials Selection for Engineering Design*; Prentice Hall International: Upper Saddle River, NJ, USA, 1999; ISBN 978-0132425957.
4. Shackelford, J.F. *Introduction for Materials Science for Engineers*, 4th ed.; Prentice-Hall International: Upper Saddle River, NJ, USA, 1998; ISBN 978-0130112873.
5. Smith, W.F.; Hashemi, J. *Foundations of Materials Science and Engineering*, 5th ed.; McGraw-Hill Education: New York, NY, USA, 2009.
6. Hsi, S.; Linn, M.C.; Bell, J.E. The role of spatial reasoning in engineering and the design of spatial instruction. *J. Eng. Educ.* **1997**, *86*, 151–158. [CrossRef]
7. Garmendia, M.; Guisasola, J.; Sierra, E. First-year engineering students' difficulties in visualization and drawing tasks. *Eur. J. Eng. Educ.* **2007**, *32*, 315–323. [CrossRef]
8. Vergara, D.; Rubio, M.P.; Lorenzo, M. New computer teaching tool for improving students´ spatial abilities in continuum mechanics. *IEEE Technol. Eng. Educ.* **2012**, *7*, 44–48.
9. Meagher, K.A.; Doblack, B.N.; Ramirez, M.; Davila, L.P. Scalable nanohelices for predictive studies and enhanced 3D visualization. *J. Vis. Exp.* **2014**, *93*, e51372. [CrossRef] [PubMed]
10. Vergara, D.; Rubio, M.P.; Lorenzo, M. A virtual environment for enhancing the understanding of ternary phase diagrams. *J. Mater. Educ.* **2015**, *37*, 93–101.
11. Rafi, A.; Khairul, A.; Samad, A.; Maizatul, H.; Mahadzir, M. Improving spatial ability using a web-based virtual environment (WbVE). *Autom. Const.* **2005**, *14*, 707–715. [CrossRef]
12. Vergara, D.; Lorenzo, M.; Rubio, M.P. Virtual environments in materials science and engineering: The students' opinion. In *Handbook of Research on Recent Developments in Materials Science and Corrosion Engineering Education*, 1st ed.; Lim, H., Ed.; IGI Global: Hershey, PA, USA, 2015; pp. 148–165.
13. Öz, C.; Serttaş, S.; Ayar, K.; Fındık, F. Effect of virtual welding simulator on TIG welding training. *J. Mater. Educ.* **2015**, *37*, 197–218.
14. Wang, P.; Wu, P.; Wang, J.; Chi, H.-L.; Wang, X. A critical review of the use of virtual reality in construction engineering education and training. *Int. J. Environ. Res. Public Health* **2018**, *15*, 1204. [CrossRef] [PubMed]
15. Sorby, S.A.; Baatmans, B.J. The development and assessment of a course for enhancing the 3-D spatial visualization skills of first year engineering students. *J. Eng. Educ.* **2000**, 301–307. [CrossRef]
16. Crown, S.W. Improving visualization skills of engineering graphics students using simple javascript web based games. *J. Eng. Educ.* **2001**, 347–355. [CrossRef]
17. Leopold, C.; Górska, R.A.; Sorby, S.A. International experiences in developing the spatial visualization abilities of engineering students. *J. Geom. Graph.* **2001**, *15*, 271–298.
18. Rafi, A.; Samsudin, K.A.; Ismail, A. On improving spatial ability through computer-mediated engineering drawing instruction. *Educ. Technol. Soc.* **2006**, *9*, 149–159.
19. Carbonell-Carrera, C.; Hess, M. Spatial orientation skill improvement with geospatial applications: Report of a multi-year study. *ISPRS Int. J. Geo-Inf.* **2017**, *6*, 278. [CrossRef]
20. Buckley, J.; Seery, N.; Canty, D. A heuristic framework of spatial ability: A review and synthesis of spatial factor literature to support its translation into STEM education. *Educ. Psychol. Rev.* **2018**, 1–26. [CrossRef]
21. Blatov, V.A.; Shevchenko, A.P.; Proserpio, D.M. Applied topological analysis of crystal structures with the program package ToposPro. *Cryst. Growth Des.* **2014**, *14*, 3576–3586. [CrossRef]
22. Arribas, V.; Casas, L.; Estopa, E.; Labrador, M. Interactive PDF files with embedded 3D designs as support material to study the 32 crystallographic point groups. *Comput. Geosci.* **2014**, *62*, 53–61. [CrossRef]

23. Casas, L.; Estop, E. Virtual and printed 3d models for teaching crystal symmetry and point groups. *J. Chem. Educ.* **2015**, *92*, 1338–1343. [CrossRef]
24. Sancho, E.; Araújo, E.; Conde, W.; Saraiva, I.; Maia, R.C.; Oliveira, L.M.; Sales, J.; Sombra, A.S.; Albuquerque, J.S. Models manufacturing of crystal systems: an introduction to ceramic materials. *Mater. Sci. Forum* **2018**, *912*, 280–284. [CrossRef]
25. Stukowski, A. OVITO: Open Visualization Tool. Available online: http://www.ovito.org/ (accessed on 19 September 2018).
26. Chan, C.; Fok, W. Evaluating learning experiences in virtual laboratory training through student perceptions: A case study in electrical and electronic engineering at the University of Hong Kong. *Engl. Educ.* **2009**, *4*, 70–75. [CrossRef]
27. Smith, W.F. *Ciencia e Ingeniería de Materiales*; Mc Graw Hill: New York, NY, USA, 2004; ISBN 84-481-2956-3.
28. Zander, S.; Wetzel, S.; Bertel, S. *Rotate it!* Effects of touch-based gestures on elementary school students' solving of mental rotation tasks. *Comput. Educ.* **2016**, *103*, 158–169. [CrossRef]
29. Courcel, M.J.; García, A.; Rodríguez, A.; Romero, M.A. What do university students think about the new active methodologies of education? *Revista de Curriculum y Formación del Profesorado* **2009**, *13*, 305–319.
30. León, M.J.; Crisol, E. Questionnaire design (OPPUMAUGR y OPEUMAUGR): The views and perceptions of teachers and students on the use of actives methodologies at the university. *Revista de Curriculum y Formación del Profesorado* **2011**, *13*, 305–319.
31. Smith, K.A. Cooperative learning: Making "groupwork" work. *Dir. Teach. Learn.* **1996**, *67*, 71–82. [CrossRef]
32. Hwang, W.-Y.; Hu, S.-S. Analysis of peer learning behaviors using multiple representations in virtual reality and their impacts on geometry problem solving. *Comput. Educ.* **2013**, *62*, 308–319. [CrossRef]
33. Conradty, C.; Bogner, F.X. Hypertext or textbook: Effects on motivation and gain in knowledge. *Educ. Sci.* **2016**, *6*, 29. [CrossRef]
34. LaForce, M.; Noble, E.; Blackwell, C. Problem-based learning (PBL) and student interest in stem careers: The roles of motivation and ability beliefs. *Educ. Sci.* **2017**, *7*, 92. [CrossRef]
35. Arango-López, J.; Cerón, C.C.; Collazos, C.A.; Gutiérrez, F.L.; Moreira, F. CREANDO: Tool for creating pervasive games to increase the learning motivation in higher education students. *Telemat. Inform.* **2018**, in press.
36. Vergara, D.; Rubio, M.P.; Lorenzo, M. Interactive virtual platform for simulating a concrete compression test. *Key Eng. Mater.* **2014**, *572*, 582–585. [CrossRef]
37. Vergara, D.; Rubio, M.P. The application of didactic virtual tools in the instruction of industrial radiography. *J. Mater. Educ.* **2015**, *37*, 17–26.
38. Vergara, D.; Rubio, M.P.; Prieto, F.; Lorenzo, M. Enhancing the teaching/learning of materials mechanical characterization by using virtual reality. *J. Mater. Educ.* **2016**, *38*, 63–74.
39. Vergara, D.; Rubio, M.P.; Lorenzo, M. New approach for the teaching of concrete compression tests in large groups of engineering students. *J. Prof. Issues Eng. Educ. Pract.* **2017**, *143*. [CrossRef]
40. Vergara, D.; Rubio, M.P.; Lorenzo, M. On the design of virtual reality learning environments in engineering. *Multimodal Technol. Interact.* **2017**, *1*, 11. [CrossRef]
41. Chasanidou, D. Design for motivation: Evaluation of a design tool. *Multimodal Technol. Interact.* **2018**, *2*, 6. [CrossRef]

education
sciences

Article

On the Use of PDF-3D to Overcome Spatial Visualization Difficulties Linked with Ternary Phase Diagrams

Diego Vergara [1],*, Manuel Pablo Rubio [2] and Miguel Lorenzo [3]

[1]   Technological Department, Catholic University of Ávila, 05005 Avila, Spain
[2]   Construction Department, University of Salamanca, 49029 Zamora, Spain; mprc@usal.es
[3]   Department of Mechanical Engineering, University of Salamanca, 37700 Béjar, Salamanca, Spain;
      mlorenzo@usal.es
*   Correspondence: diego.vergara@ucavila.es or dvergara@usal.es; Tel.: +34-920-251-020

Received: 29 January 2019; Accepted: 25 March 2019; Published: 27 March 2019

Abstract: Despite the interesting applications that the PDF-3D offers in teaching, especially for subjects related to spatial comprehension difficulties, such a didactic tool is not well known in the education sector. Thus, a proposal of using PDF-3D in engineering studies is presented in this paper, specifically, in the field of teaching ternary phase diagrams (TPDs). The didactic resource—easy to design and easy to use—allows students to overcome spatial visualization difficulties linked with TPDs. According to students' opinions, the PDF-3D is an effective tool to use in any topic related to spatial difficulties and, in addition, is a friendly and easy-to-use tool. This fact and the simplicity of designing a PDF-3D make it a useful tool for educational aims.

Keywords: ternary phase diagrams; spatial visualization; PDF-3D; engineering education

1. Introduction

Since engineering studies are closely related to spatial concepts in many subjects [1–5] and it is proved that spatial ability skills can be improved through training [6–12], instructors are continuously developing both new educational methods and new tools for enhancing the students' spatial skills [13–18]. One of the topics covered in the study of materials science and engineering is ternary phase diagrams (TPDs) [19–21], for which students exhibit spatial comprehension difficulties [22]. Thus, several papers deal with this topic, looking for a solution by developing their own virtual and computational applications [23–27]. Several of these applications are designed for professional or research activities and, consequently, a wider knowledge is required to use them [23,26,27]. In such cases, designing these virtual applications is not a simple task, and a lot of time and specific knowledge of programming is required [23,27]. Additionally, special tools are needed, e.g., stereo display methods [26]. On the other hand, other applications are mainly focused on education [22,24,25], i.e., they try to help students to qualitatively understand the concepts of TPDs (without considering the equations representing curves in a TPD). This way, this type of application is easy for both instructors (easy to design) and students (easy to use).

In recent works [15,16], one of the methods to overcome spatial comprehension difficulties in engineering students involves the use of PDF-3D. Unfortunately, the use of this tool in the engineering education field is still unexploited, despite its clear advantages in the teaching–learning process; in fact, it is easy to design, easy to use, has good accessibility, and does not require any special software or hardware. Furthermore, this tool has not yet been applied in the teaching of TPDs—as far as the authors knows. The aim of this paper is to fill this gap, by analyzing the applicability of PDF-3D in the teaching–learning process of TPDs. A design scheme that facilitates the reproducibility of a PDF-3D is presented in this paper in order to help other educational staff to reproduce this type of didactic resource. Thus, since PDF-3D is both easy to design and easy to use, this paper shows the potential application of PDF-3D in the educational field for subjects requiring spatial skills.

2. Ternary Phase Diagrams in Education

Since phase transformations are important in many industrial processes involving heat treatment of metal alloys, the educational planning of both mechanical and materials engineering studies must include instruction on phase diagrams and, consequently, their associated microstructures. However, most of the educational textbooks only focus on binary phase diagrams (BPDs), which can be represented by a simple 2D plot, and they do not include, or provide scarcely any content on TPDs. Commonly, TPDs are represented by complex 3D plots with a triangular base, placing at the edges of the triangle the three elements of the metal alloy. TPDs are especially useful in the industrial field, since they help to analyze the microstructure (and, indirectly, the mechanical properties) of innumerable metal alloys composed of a mixture of three elements (e.g., Mo–Si–B), as well as ceramic and inorganic glass systems composed of three compounds (e.g., $NaO–CaO–SiO_2$). Because of the spatial visualization difficulties of the 3D TPDs, which can be especially complex in certain cases, a common technique used is the analysis of isothermal sections of TPDs (several online explanations can be found on how a given composition of elements A, B, C ($x\%$ A, $y\%$ B, and $z\%$ C) is identified or created in a TPD [28,29]). Furthermore, the TPD is simplified sometimes to a BPD, just considering an isoconcentration section of the TPD [30].

3. PDF-3D

The PDF-3D format allows one to open, visualize, and interactively move 3D models of diverse elements (buildings, machinery parts, and 3D diagrams) that were generated by different techniques, e.g., 3D scanning, photogrammetry, and modeling. It is just a file (*.pdf) that can be opened in a conventional pdf software and allows to easily transmit tridimensional information through the internet.

3.1. PDF-3D Design Process

Taking into account the potential usefulness that this computer application has in the education field, it seems interesting to briefly explain the general process used in designing a PDF-3D. This process includes several phases where different software must be used: (1) obtaining the 3D object by scanning or modelling it (for instance, by using Autodesk 3D studio Max), (2) converting the model into an OBJ extension, (3) converting the model into a universal 3D (U3D) extension by using Adobe Photoshop, and, finally, (4) generating a PDF-3D file (Adobe Acrobat). For the sake of clarity, a scheme of the design process is shown in Figure 1.

The intermediate step is mandatory if the latest version of Adobe Acrobat (Adobe Acrobat DC) is used, since such a version only admits the file format "U3D". However, previous versions of Adobe Acrobat allow to directly import models from the CAD modelling software. On the other hand, also other types of commercial software, such as "PDF3D ReportGen", "Tretra4D Converter", or "3D PDF Maker" allow direct importation from any CAD modelling software. However, the main shortcomings of these software are their cost (they are expensive) and their complex programming language (they require complex learning). In the authors' opinion, the process presented here is the simplest one using conventional software.

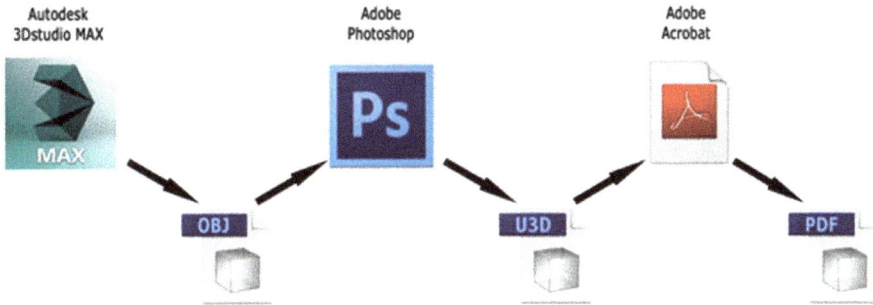

Figure 1. Design scheme of a common process for generating a PDF-3D.

3.2. PDF-3D of a TPD

In the field of mechanical and materials engineering, the design of a PDF-3D for a TPD starts by making a model of a ternary diagram with a 3D design tool, such as 3DStudio Max (Figure 2). In Figures 2–4, an example of a TPD, consisting of two binary eutectics and one binary isomorphous, is represented. The surfaces where a phase transition appears in the TPD are generated by using non-uniform rational B-spline (NURB)-type curves (), according to a mathematical model widely spread for generating and representing curves and surfaces (Figure 3). Once the TPD is modelled, it is exported as an OBJ file (a common file type for software exchanging 3D objects), and later this file is opened as a 3D model with the well-known image software Adobe Photoshop (Figure 1). The file is then exported in universal 3D format (U3D), a file format that can be read by Adobe Acrobat, just when the final step is performed, including the U3D file and generating the PDF-3D file. The final appearance of the TPD is shown in Figure 4.

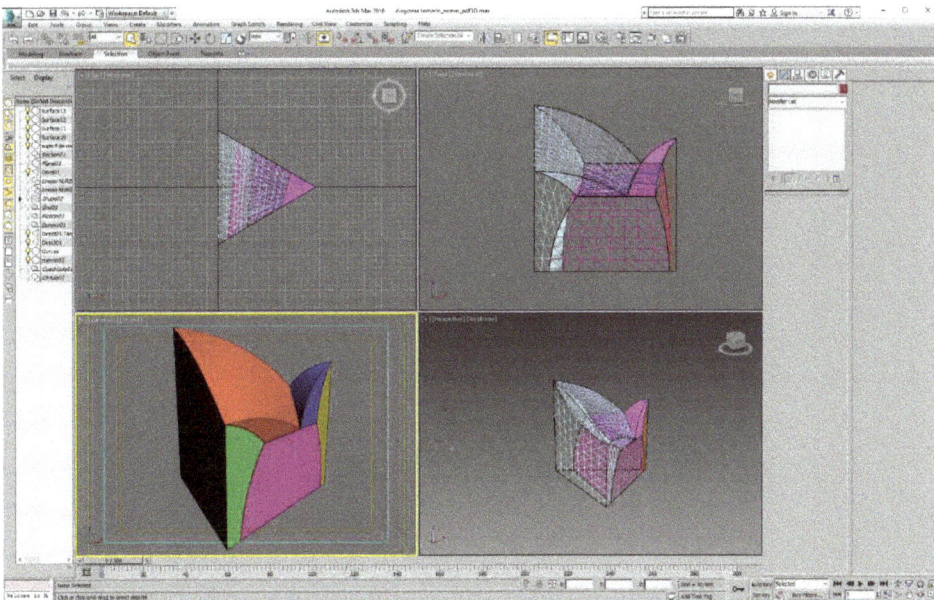

Figure 2. View of the ternary phase diagram (TPD) surfaces modelled in Autodesk 3DStudio Max.

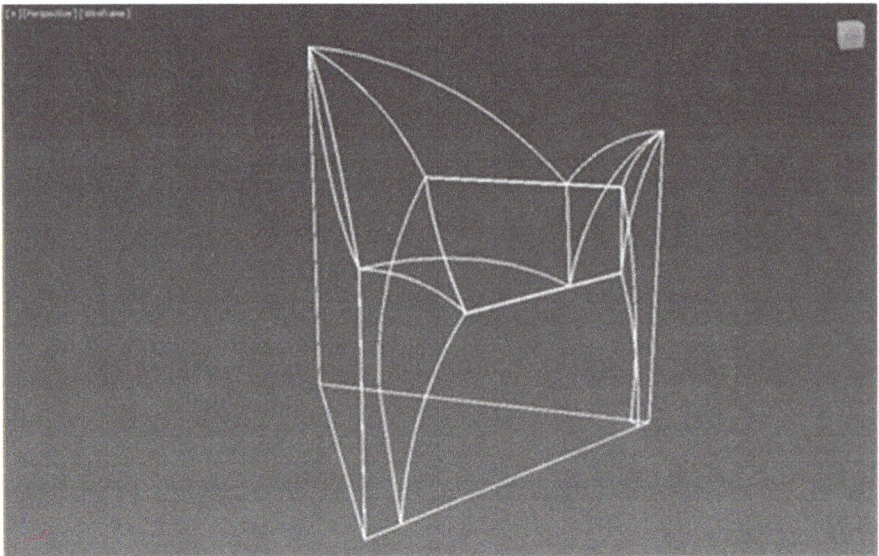

Figure 3. Non-uniform rational B-spline (NURB) curves defining a TPD.

In Figure 4, some of the PDF-3D features (useful for working with 3D objects) can be appreciated, namely, possibility of cutting and obtaining sections of the 3D object, turning and rotating the object, generating transparent zones, hiding zones, and even measuring distances in the model. Taking into account the serious spatial visualization problems in the student body, this tool is really useful for an introductory teaching of TPDs.

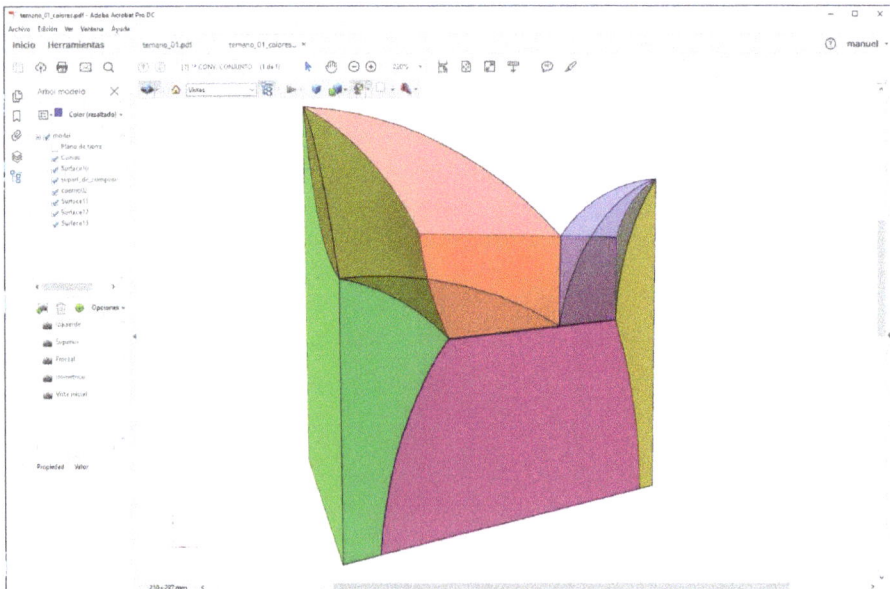

Figure 4. View of the TPD integrated in a PDF-3D.

4. Discussion

Nowadays, different options with increasing levels of complexity are available for developing an interactive computer-based tool to be used for explaining a concept involving spatial visualization skills. These tools can be classified into three groups: Group I: Applications directly programmed in a high-level language Group II: Applications created by using a specific development engine; and Group III: Applications that are obtained by using a standard software for generating 3D content, e.g., PDF-3D. For the selection of the group type to be used, the following issues must be taken into account:

- Difficulty in design and production

  - Need of scripts (thematic, technical, and artistic)
  - Need of technical knowledge
  - Required software and equipment
  - Time costs and working costs
  - Viability of the selected theme

- Degree of usability, interactivity, and complexity
- How the tool is distributed to the students

  - Unique file or a self-extracting file
  - Size of the file
  - Availability from the internet

- Need of equipment, software, and previous knowledge for using the tool

On the basis of these issues, applications included in Groups I and II exhibit a high difficulty in both design and development processes, since high-level technical knowledge is required, and a large amount of time and work must be applied [31]. Regarding the resulting tool, its size is large and, commonly, it is composed of multiple files, which makes the distribution to the students difficult. Additionally, high-performance microprocessors and graphical cards are necessary to run the tool. On the contrary, these tools allow a high level of interactivity, visual quality, concept development, and, even, content evaluation. Finally, these tools are more elaborated and controlled.

On the other hand, tools included in Group III (as in the case of PDF-3D) do not require a complex design, development processes, or a specific software. They just need the 3D model generated by a CAD software and a pdf reader software (e.g., the free Adobe Acrobat Reader), which are basic requirements for any engineering instructor. In addition to the clear advantage of the ease of design, this type of software allows an acceptable degree of interactivity and generates small-size files that can be easily distributed over the internet. The computer application PDF-3D allows to rotate and frame (to pan) the models, zooming, obtaining images, measuring, adding comments, hiding elements, selecting different points of view, modifying the type of visualization, changing the illumination, modifying the background, and making interactive sections of the model. Its main shortcomings are the low degrees of complexity, interactivity, and development. Furthermore, a commercial format developed by Adobe is required that, consequently, is limited to the Adobe specifications. Although interactivity is one of the properties most demanded by students in computer applications [32–34], the level of such a feature in Group III is enough for most of the virtual tools developed in engineering education (without professional purposes). Specifically, in the case of TPD, the PDF-3D allows students to spatially understand the different regions (3D geometry of the different phases) in an interactive way. Thus, this educational tool is a good complement to other online/interactive simulations that only explain simple 2D plots (isothermal sections of a TPD) [28,29].

4.1. Simplicity of Design (PDF-3D of a TPD)

PDF-3D could be considered as a promising tool to be used in the teaching–learning process of engineering degrees to overcome problems related to spatial visualization, which is a serious matter for engineers who must carry out tasks during their student life [4] and also during their professional careers [35,36]. The design process of a PDF-3D is simpler than the design process used in other interactive computer-based tools, and PDF-3D is sufficient for the required aims in the teaching–learning process developed in engineering studies. Specifically, in the case of TPDs, PDF-3D is adequate for developing the basic design required in mechanical engineering studies, i.e., without precise mathematical equations that define the curves. The main aim sought by an instructor during the explanation of a TPD in mechanical engineering studies is achieving the spatial understanding of how such a diagram is formed, whereas the mathematical equations defining the curves of the TPDs are not important. Thus, this type of models is sufficient for teaching basic concepts regarding TPDs and helping students understand the spatial comprehension difficulties that the use of TPDs exhibits in a professional environment.

4.2. Educational Experience

The students´ opinions after using the previously exposed didactic tool are presented in this section. To this aim, a quasi-experimental research was carried out on a total of 32 students enrolled in the Materials Science and Engineering course (in the Mechanical Engineering Degree). The survey questions were focused on, firstly, how helpful the PDF-3D tool is for students in the task of spatially understanding a TPD and, secondly, how easy it is to use such a didactic tool (Table 1).

Table 1. Questions included in the student survey.

Number	Question	Response
1	Do you think that the PDF-3D tool helps you to spatially understand the ternary phase diagrams?	Rate from 1 to 10
2	Do you think that the PDF-3D is an effective tool to use in topics related to spatial difficulties?	Rate from 1 to 10
3	How easy was it to use the PDF-3D tool of ternary phase diagrams?	Rate from 1 to 10
4	Could you indicate the positive aspects of using PDF-3D for teaching TPD?	

The obtained results revealed that the help provided by using the PDF-3D tool to spatially understand the TPDs was well rated by the students, as reflected in the average values of the answers to question 1 (9.3 out of 10 with a standard deviation of 0.24). Likewise, according to the students' answers to question 2 (Table 1), the PDF-3D is considered an effective tool to be used in other topics related to spatial difficulties (9.6 out of 10 with a standard deviation of 0.28). Furthermore, students gave the highest rating to the ease of use of the PDF-3D (9.9 out of 10 with a standard deviation of 0.20) and they highlighted other positive aspects of such a didactic tool, such as the availability to use it at home and the speed of opening the tool with a conventional computer.

It is worth noting here that the result of question 3 (9.9 out of 10) revealed how easy it is to use a PDF-3D, since mechanical engineering students generally give lower ratings to other virtual didactic tools which are easy to use [18,37,38]. This fact, together with the simplicity of designing a PDF-3D (compared with other didactic virtual tools usually designed with a more complex software), makes PDF-3D a promising application for the teaching–learning process in engineering studies.

5. Conclusions

Despite the didactic usefulness of the PDF-3D for subjects involving concepts with spatial visualization difficulties, there is a lack of research papers that show the use of this tool in the classroom. Since the PDF-3D tool is quite new, the tool and its potential applications are still unknown to the world of education. Authors consider the PDF-3D to be an educational tool with interesting applications in any subject or topic related to spatial visualization difficulties, e.g., technical drawing, crystal lattices, biochemistry.

In this paper, a practical application using a PDF-3D in engineering studies in shown in order to teach ternary phase diagrams, which are characterized by serious spatial visualization difficulties. Both students and authors are really satisfied with the possibilities that such a didactic tool offers. Furthermore, this paper includes an explanation of the design process to create a PDF-3D, which is simpler and much easier to make than other didactic virtual resources, so that anybody can reproduce this idea, adapting it to specific applications.

Author Contributions: Designing the virtual tool, D.V. and M.P.R.; software, M.P.R.; validation, D.V. and M.L.; writing—review and editing, D.V., M.P.R., and M.L.

Acknowledgments: The authors wish to acknowledge the financial support provided by the following Spanish Universities: University of Salamanca (Proyect ID2OI6/212) and Innovation Institute UFV (Grant 2017).

Conflicts of Interest: The authors declare no conflict of interest.

References

1. Saorin, J.L.; Navarro, R.E.; Martín, N.; Martín, J.; Contero, M. Spatial skills and its relationship with the engineering studies. *Dyna* **2001**, *84*, 721–732.
2. Vergara, D.; Rubio, M.P. Active methodologies through interdisciplinary teaching links: Industrial radiography and technical drawing. *J. Mater. Educ.* **2012**, *34*, 175–186.
3. Carbonell-Carrera, C.; Hess Medler, S. Spatial orientation skill improvement with geospatial applications: Report of a multi-year study. *ISPRS Int. J. Geo-Inf.* **2017**, *6*, 278. [CrossRef]
4. Ha, O.; Fang, N. Spatial ability in learning engineering mechanics: Critical review. *J. Prof. Issues Eng. Educ. Pract.* **2016**, *142*, 04015014. [CrossRef]
5. Daineko, Y.; Dmitriyev, V.; Ipalakova, M. Using virtual laboratories in teaching natural sciences: An example of physics courses in university. *Comput. Appl. Eng. Educ.* **2017**, *25*, 39–47. [CrossRef]
6. Sorby, S.A.; Baartmans, B.J. The development and assessment of a course for enhancing the 3-D spatial visualization skills of first year engineering students. *J. Eng. Educ.* **2000**, *89*, 301–307. [CrossRef]
7. Guimaraes, L.C.; Garcia, R.; Belfort, E. Tools for teaching mathematics: A case for Java and VRML. *Comput. Appl. Eng. Educ.* **2000**, *8*, 157–161. [CrossRef]
8. Leopold, C.; Górska, R.A.; Sorby, S.A. International experiences in developing the spatial visualization abilities of engineering students. *J. Geom. Graph.* **2001**, *5*, 81–91.
9. Alias, M.; Black, T.; Gray, D. Effect of instructions on spatial visualisation ability in civil engineering students. *Int. Educ. J.* **2002**, *3*, 1–12.
10. Rafi, A.; Khairul, A.; Samad, A.; Maizatul, H.; Mahadzir, M. Improving spatial ability using a web-based virtual environment (WbVE). *Autom. Constr.* **2005**, *14*, 707–715. [CrossRef]
11. Rafi, A.; Samsudin, K.A.; Ismail, A. On improving spatial ability through computer-mediated engineering drawing instruction. *Educ. Technol. Soc.* **2006**, *9*, 149–159.
12. Prieto, G.; Velasco, A. Does spatial visualization ability improve after studying technical drawing? *Qual. Quant.* **2010**, *44*, 1015–1024. [CrossRef]
13. Martínez, E.; Carbonell, V.; Florez, M.; Amaya, J. Simulations as a new physics teaching tool. *Comput. Appl. Eng. Educ.* **2010**, *18*, 757–761. [CrossRef]
14. Martín-Gutiérrez, J.; García-Domínguez, M.; Roca-González, C.; Sanjuán-Hernanpérez, A.; Mato-Carrodeguas, C. Comparative analysis between training tools in spatial skills for engineering graphics students based in virtual reality, augmented reality and PDF3D technologies. *Procedia Comput. Sci.* **2013**, *25*, 360–363. [CrossRef]
15. Martín-Gutiérrez, J.; García-Domínguez, M.; Roca-González, C. Using 3D virtual technologies to train spatial skills in engineering. *Int. J. Eng. Educ.* **2015**, *31*, 323–334.
16. Melgosa, C.; Ramos, B.; Baños, M.E. Interactive learning management system to develop spatial visualization abilities. *Comput. Appl. Eng. Educ.* **2015**, *23*, 203–216. [CrossRef]
17. Šafhalter, A.; Bakracevic, K.; Glodež, S. The effect of 3D-modeling training on students' spatial reasoning relative to gender and grade. *J. Educ. Comput. Res.* **2016**, *54*, 395–406. [CrossRef]

18. Vergara, D.; Rubio, M.P.; Lorenzo, M. A virtual resource for enhancing the spatial comprehension of crystal lattices. *Educ. Sci.* **2018**, *8*, 153. [CrossRef]
19. Browtow, W. Instruction in materials science and engineering: Modern technology and the new role of the teacher. *Mater. Sci. Eng. A* **2001**, *302*, 181–185. [CrossRef]
20. Mansbach, R.; Ferguson, A.; Kilian, K.; Krogstad, J.; Leal, C.; Schleife, A.; Trinkle, D.R.; West, M.; Herman, G.L. Reforming an undergraduate materials science curriculum with computational modules. *J. Mater. Educ.* **2016**, *38*, 161–174.
21. Pfennig, A.; Hadwiger, P. Peer-to-peer lecture films—A successful study concept for a first-year laboratory material science course. *Procedia Soc. Behav. Sci.* **2016**, *228*, 24–31. [CrossRef]
22. Vergara, D.; Rubio, M.P.; Lorenzo, M. New virtual application for improving the students' understanding of ternary phase diagrams. *Key Eng. Mater.* **2014**, *572*, 578–581. [CrossRef]
23. Tomiska, J. "ExTherm 2": An interactive support package of experimental and computational thermodynamics. *Calphad: Comput. Coupling Phase Diagr. Thermochem.* **2009**, *33*, 288–294. [CrossRef]
24. Glasser, L.; Herráez, A.; Hanson, R.M. Interactive 3D phase diagrams using Jmol. *J. Chem. Educ.* **2009**, *86*, 566. [CrossRef]
25. Vergara, D.; Rubio, M.P.; Lorenzo, M. A virtual environment for enhancing the understanding of ternary phase diagrams. *J. Mater. Educ.* **2015**, *37*, 93–102.
26. Kang, J.; Liu, B. Stereo 3D spatial phase diagrams. *J. Alloy. Compd.* **2016**, *673*, 309–313. [CrossRef]
27. Mu, Y.; Bao, H. A three-dimensional topological model of ternary phase diagram. *J. Phys. Conf. Ser.* **2017**, *787*, 012005. [CrossRef]
28. Using a Triangular (Ternary) Phase Diagram. Available online: https://www.youtube.com/watch?v=gGYHXhcKM5s (accessed on 23 March 2019).
29. Ternary Phase Diagram Basics (Interactive Simulation). Available online: https://www.youtube.com/watch?v=GXap5CC8MN4 (accessed on 23 March 2019).
30. West, D.R.F. *Ternary Equilibrium Diagrams*; Chapman & Hall: New York, NY, USA, 1982.
31. Vergara, D.; Rubio, M.P.; Lorenzo, M. On the design of virtual reality learning environments in engineering. *Multimodal Technol. Interact.* **2017**, *1*, 11. [CrossRef]
32. Chan, C.; Fok, W. Evaluating learning experiences in virtual laboratory training through student perceptions: A case study in electrical and electronic engineering at the University of Hong Kong. *Engl. Educ.* **2009**, *4*, 70–75. [CrossRef]
33. Vergara, D.; Rubio, M.P. The application of didactic virtual tools in the instruction of industrial radiography. *J. Mater. Educ.* **2015**, *37*, 17–26.
34. Vergara, D.; Rubio, M.P.; Lorenzo, M. New approach for the teaching of concrete compression tests in large groups of engineering students. *J. Prof. Issues Eng. Educ. Pract.* **2017**, *143*, 05016009. [CrossRef]
35. Hsi, S.; Linn, M.C.; Bell, J.E. The role of spatial reasoning in engineering and the design of spatial instruction. *J. Eng. Educ.* **1997**, *86*, 151–158. [CrossRef]
36. Baronio, G.; Motyl, B.; Paderno, D. Technical drawing learning tool-level 2: An interactive self-learning tool for teaching manufacturing dimensioning. *Comput. Appl. Eng. Educ.* **2016**, *24*, 519–528. [CrossRef]
37. Vergara, D.; Rubio, M.P.; Prieto, F.; Lorenzo, M. Enhancing the teaching/learning of materials mechanical characterization by using virtual reality. *J. Mater. Educ.* **2016**, *38*, 63–74.
38. Vergara, D.; Rodríguez-Martín, M.; Rubio, M.P.; Ferrer, J.; Núñez, F.J.; Moralejo, L. Technical staff training in ultrasonic non-destructive testing using virtual reality. *Dyna* **2018**, *94*, 150–154.

education
sciences

MDPI

Article

Solving Power Balance Problems in Single-Traction Tractors Using PTractor Plus 1.1, a Possible Learning Aid for Students of Agricultural Engineering

Marta Gómez-Galán, Ángel Carreño-Ortega, Javier López-Martínez and Ángel-Jesús Callejón-Ferre *

Department of Engineering, University of Almería, Agrifood Campus of International Excellence (CeiA3), s/n, La Cañada, Almería 04120, Spain; mgg492@ual.es (M.G.-G.); acarre@ual.es (Á.C.-O.); jlm167@ual.es (J.L.-M.)
* Correspondence: acallejo@ual.es

Received: 4 April 2018; Accepted: 6 May 2018; Published: 8 May 2018

Abstract: Tractors are used to perform jobs that require different types of agricultural tools to be attached to their rear, to their front, or both. These tools may need to be dragged, towed, or suspended above ground, and sometimes require a power supply; this is usually obtained via a hydraulic system or from the tractor's power take-off system. When tractors have to work with such tools on different types of soils and on different slopes, the need arises to calculate the power the tractor engine will have to produce. In the classroom, this is normally calculated manually with the help of a calculator. This work, however, describes a computer program (written in Delphi and operating under Windows) that rapidly solves the most common types of power balance problems associated with single-traction tractors. The value of this software as a learning aid for students of agricultural engineering is discussed.

Keywords: improving classroom teaching; simulations; teaching/learning strategies

---

1. Introduction

The use of new technologies in teaching has, currently, a very important place in the field of education [1]. Thanks to its use, it is possible to extend teaching to students regardless of their level of knowledge, schedule availability, location, or other limitations [2].

Agricultural Engineering university studies are undergoing great changes and improvements, thanks to the use of these technologies. Thus, it has been demonstrated the improvement in the development of agricultural engineering class sessions by using the flipped classroom learning environments (combination of the conventional and virtual teaching) [3]. Other authors proposed the implementation of distance-learning in engineering studies and use it for the benefit of increasing their attractiveness [4].

Among the important advances that have been made in the teaching of engineering and the use of new technologies [5], the following stand out: the tools for simulation of crop growing in agricultural [6], the use of devices to support the crop growing, advising, for example, the need of irrigation or fertilization [7,8], the use of supervisory control and data acquisition systems (SCADA) for the installation design, as drip irrigation systems [9], the use of software of finite element analysis (FEA) for the simulation of structural components of vehicles and machinery [10], the implementation of collaborative computer-aided engineering (CAE) for collaborative tools that allow the integration of different engineering specialties [11], the use of interactive exercises (using software), the use of information in real time, the use, by students, of graphics software for modelling or a virtual reality system for engineering laboratory education [12,13], the combining of attending theory classes with web-based learning and the use of software for solving problems (even in examination situations) [14,15], the use of tools for videoconferences between professors and students, also allowing group classes online [16,17], the use of augmented reality in order to improve the learning results [18,19], and the use of IT tools in the laboratory as a learning aid [20]. In addition, the use of web-based technologies, such as Moodle, as a support to traditional teaching [21,22].

In agricultural engineering, a main aspect is the knowledge of the machinery and equipment used [23]. The tractor is a vehicle widely used in several tasks. The technological development has also contributed to numerous advances in this type of agricultural vehicle [24], such as, for example, improvements in steering systems [25], the development of hybrid tractors [26], the use of algorithms and simulation to improve the suspension systems [27], or the use of a sensor to control the wear of some elements [28].

Therefore, during university studies in this specialty, students must acquire the knowledge for the design and calculation of the main parameters related with the tractors. One of these parameters is the engine power of an agricultural tractor [29]. One way to obtain the torque and power values of a tractor's engine is through the use of a dynamometer [30]. Also, some software solutions have been developed for these and other types of calculations [31], and mobile applications, which are a complement to the study of tractors [32].

This work describes the development and use of software that students could use to check their manually-produced solutions to problems on the behaviour of single-traction tractors in the field. Tractors are used to perform jobs that require different types of agricultural tools to be attached to their rear, to their front, or both. These tools may need to be dragged, towed, or suspended above ground, and sometimes require a power supply; this is usually obtained via a hydraulic system or from the tractor's power take-off system [33]. When tractors are working with such tools on different types of soils and on different slopes, the problem arises of calculating the power the tractor engine needs to produce [33,34]. Traditionally, such problems are tackled using the nomenclature described in Table 1 and the equations shown in Figure 1.

This calculation, therefore, demands knowledge of certain physical characteristics of the tractor–tool–soil system, and the resistance to movement caused by the weight supported by each of the tractor's wheels. The software developed takes all these variables into account, and can determine the power required of the engine under a wide range of work conditions. The value of this software as a learning aid for students of agricultural engineering is discussed.

Table 1. Nomenclature used.

Symbol	Magnitude	Symbol	Magnitude
N_e	Effective power of the engine	p	Pressure in the hydraulic system
N_T	Power transmitted	q	Hydraulic flow
η_m	Mechanical yield of the transmission	η_g	Overall pump performance
$N_{s.e.}$	Power consumed by the electrical system	T	Moment of the power take-off system
N_h	Power consumed by the hydraulic system	ω	Velocity of power take-off
$N_{t.d.f.}$	Power supplied by the power take-off system	U	Peripheral forces in drive wheels
N_U	Usable power	R	Resistance to movement
N_ρ	Power losses in movement	V_t	Theoretical velocity
N_α	Power associated with the slope	V_R	True velocity
N_a	Power associated with acceleration	P_T	Cross-sectional weight of the system
N_σ	Power lost due to sliding	F_i	Inertia of acceleration
N_Z	Power transferred to the towing bar	Z	Force associated with the towing bar
V	Volt	I	Ampere
η_b	Bar yield (Z)		

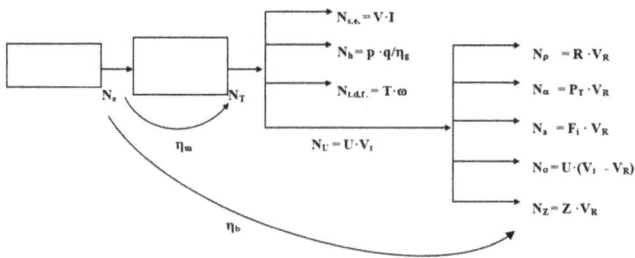

Figure 1. Power balance for a single-traction tractor.

2. Materials and Methods

A series of algorithms was produced that contemplates the majority of power-requirement situations encountered in the normal use of single-traction tractors. These situations were catalogued by examining the questions given in exercises and examinations to students of agricultural engineering at the University of Almería, Spain, in recent years. These algorithms were incorporated into a program designed to calculate the power required of the tractor engine in different scenarios. The program was written in Delphi [35], a high level language, and compiled to generate a file executable under the Windows operating system. The flowchart of the program is showed in Figure 2, where each step of the diagram is explained below in this section.

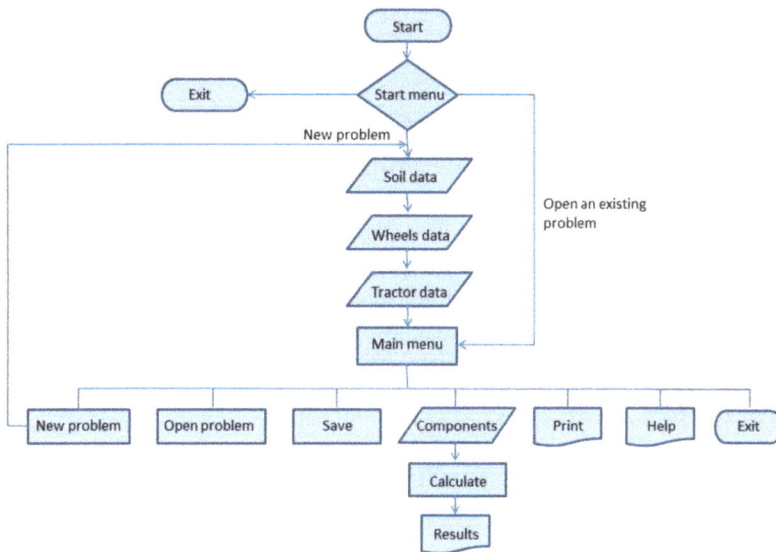

Figure 2. Program flowchart.

2.1. Start-Up Window

On starting the program, a window (Figure 3) appears that allows the user to choose between two options: to begin work on a new problem, or to re-open a saved problem. If desired, the program can be exited at this point by pressing the EXIT button.

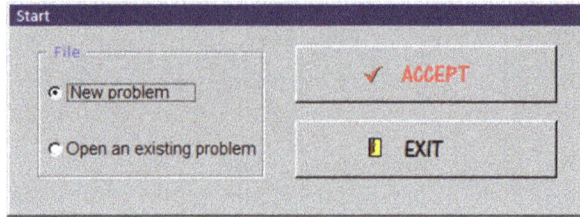

Figure 3. Start-up window.

2.2. Soil Data

If the "new problem" option is chosen, a new window appears (Figure 4) for the incorporation of data pertaining to the soil over which the tractor must move (maximum specific resistance of the soil and slope). If these data are not required for the solution of the problem, values of 0 are entered. When all data have been entered, the ACCEPT button is pushed.

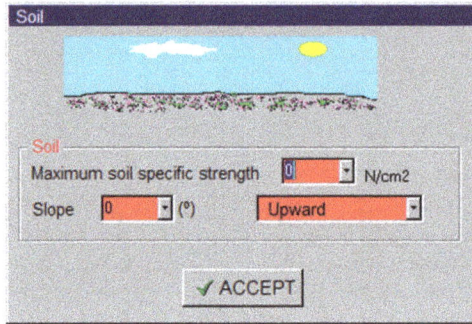

Figure 4. Soil data input window.

2.3. Data Pertaining to the Forward and Rear Wheels

Once the soil data are entered, a new window (Figure 5) appears for the introduction of data pertaining to the front wheels of the tractor (forward ballast, the coefficient of rolling resistance of the front wheels, and the front wheel radius). When this has been done, the ACCEPT button is pushed. A new window, similar to the last one, then opens for the rear wheel data to be introduced. The operator then proceeds as above.

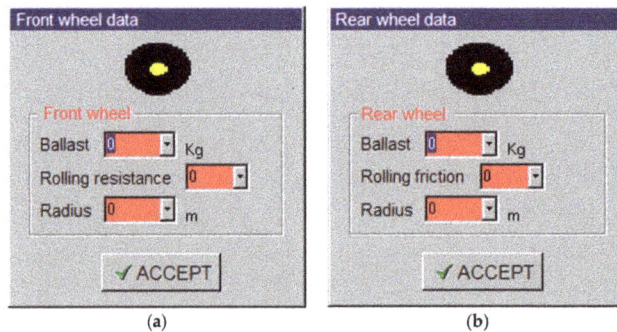

Figure 5. Tractor wheel data input window; (a) Front wheel; (b) Rear wheel.

2.4. Chassis and Engine Data

Once the above data have all been introduced, a new window (Figure 6) opens to allow the introduction of the chassis and engine data; this includes the distance between the axels, the tractor's centre of gravity, its weight, the power loss in the hydraulic system, the power reserve of the engine, the mechanical yield of the transmission, power loss in the electrical system, the theoretical velocity of the tractor, the true velocity of the tractor, and coefficient of sliding friction. In the velocity section of the window, two or three of the requested values should be introduced. A value of 0 is introduced if the true value is unknown. The CALCULATE button is then pushed for the unknown value to be determined from the data provided. The ACCEPT button is then pushed. If the values introduced are clearly incorrect (e.g., a true velocity higher than the theoretical velocity has been entered) a message to this effect will appear. Any errors can then be corrected before pushing the ACCEPT button again.

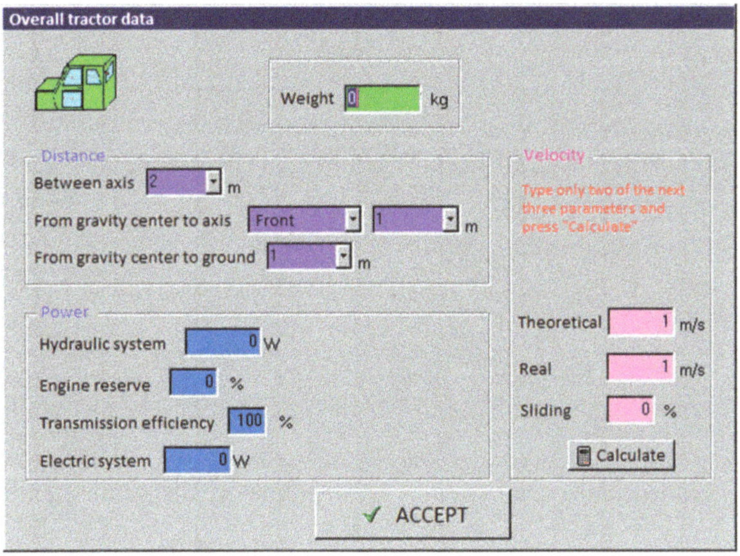

Figure 6. Chassis and engine data input window.

2.5. Main Window

When the above step has been completed, a new window (Figure 7) appears. This allows the machinery (e.g., fertilizer spreader, drill, pulveriser, mould board plough, trailer, etc.) to be connected to the tractor (either at the front or rear) to be introduced.

A desired tool is selected from the options by clicking on it and dragging to the picture of the tractor. The virtual tractor can take a maximum of one suspended or dragged tool and two forward tools; no more can be added, but choices can be changed. To eliminate a tool, all that is required is to drag it from virtual tractor and drop it in the recycle bin. The tool selection options will then become active again. The data for the attached tool can then be introduced by clicking on the "TRACTOR COMPONENTS" dropbox and choosing the tool selected; a data input window will then appear. These data can be modified by performing the process again at any time.

Figure 7. Main window.

2.6. Visualizing the Results

To see the results (Figure 8), the user only needs to do the following:

(a) Press the CALCULATE button in the main window. A window will then appear with all the results.

(b) Clicking on a result will cause a window to appear with the formula used to obtain that results. To move between the different types of result, the user need only click on the different tabs.

(c) To return to the main window, the RETURN button is pushed.

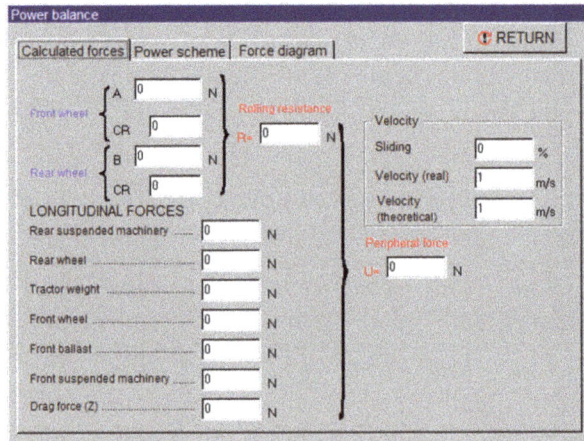

(a)

Figure 8. *Cont.*

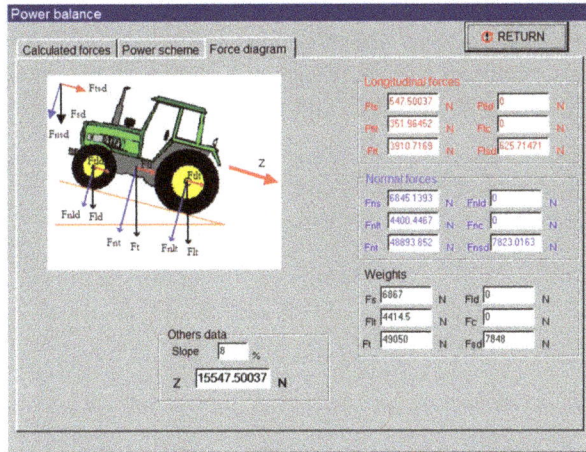

(b)

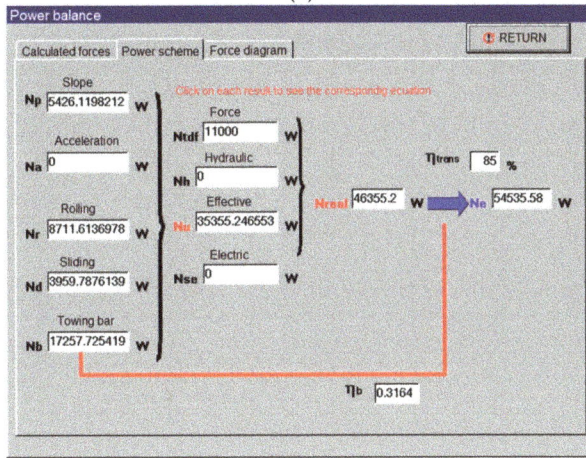

(c)

Figure 8. Results windows; (**a**) Calculated forces; (**b**) Force diagram; (**c**) Power scheme.

2.7. Saving a Problem

Being able to save a problem means the user need not put in all the data every time he or she wishes to return to it. Saving requires the user to:

Push the SAVE option in the main window.

Choose a file name for the problem and press SAVE in the corresponding window.

If a filename is chosen that already exists, a message appears asking the user if the intention is to overwrite that file. The YES button should be chosen if this is the case, the NO button if it is not. The latter choice will allow the user to provide a new name for the problem to be saved.

2.8. Recovering a Saved Problem

If the program has just been opened, a window will appear asking whether the user would like to re-open a saved problem. If this is the case, the user should press ACCEPT, and choose the problem desired.

If the program is already running and the user wants to access a saved problem, the OPEN button in the main window should be clicked. The directory and appropriate file should be sought, and the problem opened by pressing the OPEN button. This will return the user to the main window, where all the data pertaining to that problem will appear in the correct place.

2.9. Printing Results

To print the results, the user should:
Press the PRINT button in the main window.

2.10. Evaluation of the Software

To evaluate the software developed, a survey has been carried out that includes eleven questions to be answered by a group of students (Table 2). Two questions are about the characteristics of the student and the other nine questions about the use of the software. A total of sixteen students of the Grade in Agricultural Engineering of the University of Almería (Spain) have carried out the survey.

Table 2. Qualitative variables collected for each student.

Variable	Category	Abbreviation
Students		
1. Sex	Male Female	Mal Fem
2. Age	≤20 years old ≥21 years old	T1 T2
Software		
3. Is the interface easy to use?	Yes No Ever	q3y q3n q3e
4. Would it improve the interface (titles, windows, menus, icons, buttons or figures)?	Yes No Ever	q4y q4n q4e
5. Do you find useful the software to solve problems of power balances in tractors?	Yes No Ever	q5y q5n q5e
6. Is the level of knowledge required to use the software in accordance with the academic level?	Yes No Ever	q6y q6n q6e
7. Is the speed of response of the software correct? (user interaction and program)	Yes No Ever	q7y q7n q7e
8. Do you think necessary previous explanations of the teacher in the classroom for the use of the software?	Yes No Ever	q8y q8n q8e
9. Is it easy to navigate the program?	Yes No Ever	q9y q9n q9e
10. Is the information correct and updated?	Yes No Ever	q10y q10n q10e
11. Rate the software PTractor Plus 1.1. from 1 to 10:	<5 5–7 >7	L M H

The results of the survey have been statistically evaluated using the XLSTAT2018 software (Addinsoft, Paris, France).

3. Results and Discussion

The program developed, known as PTractor 1.1, quickly performs the tedious tasks required in the calculation of power balances in single-traction tractors. It could be used by engineers to rapidly assess the possibility of a tractor being able to perform a desired task, or to simulate small field trials, such as those performed with single-traction tractors by Evans, Clark & Manor [36], Shibusawa & Sasao [37], Sharma & Pandey [38]. and Raheman & Jha [39]. Perhaps more importantly, however, it has the potential to be used as a learning aid.

Generally, students in our department are first exposed to the basic concepts of the equations required to calculate power balances. They then have to use this new knowledge to work out some general problems. These calculations are performed manually with the aid of a calculator. Each question demands considerable time on the part of the student, who has to repeat the majority of the calculations made if any condition of the problem is changed. However, our students are allowed access to the program described, either individually or in groups. They can then check their calculations and easily modify variables to see what effect this has on the final result—something that should reinforce their understanding of the basic concepts they have been taught.

This type of software-supported learning for the verification of results has been tried in other areas of engineering by other authors [6,11].

Regarding the survey carried out for the students (Table 2), the statistical results obtained are showed in Table 3. Most students who have completed the questionnaires are under 21 (56.25%—T1). This means that almost half (43.75%—T2) are students who have repeated the course. Besides, male students are about twice that female students.

Table 3. Frequencies and modes for the qualitative variables by category.

Variable	Categories	Frequencies	%
Age	T1 *	9	56.25
	T2	7	43.75
Sex	Fem	5	31.25
	Mal *	11	68.75
q3	q3e	1	6.25
	q3y *	15	93.75
q4	q4e	5	31.25
	q4n	4	25.00
	q4y *	7	43.75
q5	q5e	1	6.25
	q5y *	15	93.75
q6	q6e	3	18.75
	q6n	3	18.75
	q6y *	10	62.50
q7	q7e	1	6.25
	q7y *	15	93.75
q8	q8e *	7	43.75
	q8n	4	25.00
	q8y	5	31.25
q9	q9e	5	31.25
	q9y *	11	68.75
q10	q10e	3	18.75
	q10y *	13	81.25
q11	H *	7	43.75
	L	2	12.50
	M *	7	43.75

* Mode.

With respect the question about the software, the 93.75% of the users consider that the software is easy to use ("q3") and fast ("q7") but, at the same time, consider that it should be improved ("q4"). This may be due to the fact that the program has been used for more than ten years without having been updated with the new Windows environments; however, almost all students find it useful ("q5").

In question 6, most students think that the software is appropriate to their knowledge; nevertheless, approximately 40% of the students think that it is not appropriate, percentage that coincides with the number of "T2" students. Also, this percentage coincides with the number of students that requires, sometimes, an explanation about the use of the software before using it ("q8").

Near 70% of the students have found the software easy to navigate ("q9") and around 80% think that the information is correct and updated ("q10"). All this is positive, but it should not lead the professors to conformism. Improvements would be necessary in the search for educational excellence. Regarding the overall evaluation of the software, the valuation has been medium-high (close to 90%; "q11").

Finally, as limitations of the software it should be noted that it does not take into account the regenerative auxiliary power of the tractors [40], nor does it solve problems of tractors with double traction. These facts, together with the updating of the interface, advise an update of the PTractor Plus software.

4. Conclusions

The software could be of use to professional engineers when power balances need to be calculated, or for undertaking simulations in the laboratory. However, it may also have great potential as a learning aid for students in real or virtual classrooms.

Also, although the students value, in general, the software positively, a short/medium term update is necessary.

Author Contributions: All authors contributed equally to the manuscript, and have approved the final manuscript.

Funding: This research received no external funding.

Acknowledgments: To Raúl Aroca-Delgado and José-Antonio López-Martínez. Also to Research Plan of the University of Almería.

Conflicts of Interest: The authors declare no conflict of interest.

References

1. Mercader, C.; Sallan, J.G. How do university teachers use digital technologies in class? *Rev. Docencia Univ.* **2017**, *15*, 257–273. [CrossRef]
2. Palkova, Z.; Vakhtina, E. Innovative learning: From multimedia to virtual worlds. In Proceedings of the 7th International Conference on Education and New Learning Technologies (EDULEARN), Barcelona, Spain, 6–8 July 2015; Gomez-Chova, L., Lopez-Martinez, A., Candel-Torres, I., Eds.; IATED: Valencia, Spain, 2015.
3. Busato, P.; Berruto, R.; Zazueta, F.S.; Silva-Lugo, J. Student performance in conventional and flipped classroom learning environments. *Appl. Eng. Agric.* **2016**, *32*, 509–518. [CrossRef]
4. Katzis, K.; Dimopoulos, C.; Meletiou-Mavrotheris, M.; Lasica, I.E. Engineering Attractiveness in the European Educational Environment: Can Distance Education Approaches Make a Difference? *Educ. Sci.* **2018**, *8*, 16. [CrossRef]
5. Babich, A.; Mavrommatis, K.T. Teaching of complex technological processes using simulations. *Int. J. Eng. Educ.* **2009**, *25*, 209–220.
6. Sanchez, J.A.; Perez, N.; Rodriguez, F.; Guzman, J.L.; Lopez, J.C. Decision support system for controlling the growth of greenhouse pepper crops base on climatic conditions. In Proceedings of the 7th Iberian Congress of Agricultural Engineering and Horticultural Sciences, Madrid, Spain, 26–29 August 2013; Tellez, F.A., Rodriguez, A.M., Sancho, I.M., Robinson, M.V., RuizAltisent, M., Ballesteros, F.R., Hernando, E.C.C., Eds.; The Universidad Politécnica de Madrid: Madrid, Spain, 2014.

7.   Perez-Castro, A.; Sanchez-Molina, J.A.; Castilla, M.; Sanchez-Moreno, J.; Moreno-Ubeda, J.C.; Magan, J.J. FertigUAL: A fertigation management app for greenhouse vegetable crops. *Agric. Water Manag.* **2017**, *183*, 186–193. [CrossRef]

8.   Marti, P.; Royuela, A. Practical sesion on the application of a robust mathematical tool for the calculation of crop water requirements in agricultural engineering. In Proceedings of the 6th International Conference on Education, Research and Innovation (ICERI), Seville, Spain, 18–20 November 2013; Chova, L.G., Martinez, A.L., Torres, I.C., Eds.; IATED: Valencia, Spain, 2013.

9.   Molina, J.M.; Ruiz-Canales, A.; Jimenez, M.; Soto, F.; Fernandez-Pacheco, D.G. SCADA Platform Combined with a Scale Model of Trickle Irrigation System for Agricultural Engineering Education. *Comput. Appl. Eng. Educ.* **2014**, *22*, 463–473. [CrossRef]

10.   Ahmad, F.; Kumar, A.; Kanwar, K.; Patil, P.P. CFX, Static Structural Analysis of Tractor Exhaust System Based on FEA. In Proceedings of the 28th International Conference on CAD/CAM, Robotics and Factories of the Future (CARs and FoF), Kolaghat, India, 6–9 January 2016; Mandal, D.K., Syan, C.S., Eds.; Springer: New Delhi, India, 2016.

11.   Sancibrian, R.; Llata, J.R.; Sarabia, E.G.; Torre-Ferrero, C.; Blanco, J.M.; San-Jose, J.T. Industry of the future: Implementation of collaborative CAE tools in industrial engineering degrees. In Proceedings of the 11th International Conference on Technology, Educarion and Development (INTED), Valencia, Spain, 6–8 March 2017; Chova, L.G., Martinez, A.L., Torres, I.C., Eds.; IATED: Valencia, Spain, 2017.

12.   Hashemipour, M.; Manesh, H.F.; Bal, M. A modular virtual reality system for engineering laboratory education. *Comput. Appl. Eng. Educ.* **2011**, *19*, 305–314. [CrossRef]

13.   Potkonjak, V.; Gardner, M.; Callaghan, V.; Mattila, P.; Guetl, C.; Petrovic, V.M.; Jovanovic, K. Virtual laboratories for education in science, technology, and engineering: A review. *Comput. Educ.* **2016**, *95*, 309–327. [CrossRef]

14.   Reimann, P.; Bull, S.; Halb, W.; Johnson, M. Design of a computer-assisted assessment system for classroom formative assessment. In Proceedings of the 14th International conference on Interactive Collaborative Learning (ICL), Piestany, Slovakia, 21–23 September 2011.

15.   Reimann, P.; Kickmeier-Rust, M.; Albert, D. Problem solving learning environments and assessment: A knowledge space theory approach. *Comput. Educ.* **2013**, *64*, 183–193. [CrossRef]

16.   Fita, A.; Monserrat, J.F.; Molto, G.; Mestre, E.M.; Rodriguez-Burruezo, A. Use of Synchronous e-Learning at University Degrees. *Comput. Appl. Eng. Educ.* **2016**, *24*, 982–993. [CrossRef]

17.   Gilarranz, C.; Olivares, J.; Munoz, M.A.; Lazaro, J.M.; Ramos-Paul, P. Collaborative tutorial groups in virtual environments with Moodle as a support. In Proceedings of the 7th Iberian Congress of Agricultural Engineering and Horticultural Sciences, Madrid, Spain, 26–29 August 2013; Tellez, F.A., Rodriguez, A.M., Sancho, I.M., Robinson, M.V., RuitzAltisent, M., Ballesteros, F.R., Hernando, E.C.C., Eds.; The Universidad Politécnica de Madrid: Madrid, Spain, 2014.

18.   Lytridis, C.; Tsinakos, A.; Kazanidis, I. ARTutor—An Augmented Reality Platform for Interactive Distance Learning. *Educ. Sci.* **2018**, *8*, 6. [CrossRef]

19.   Wei, X.D.; Weng, D.D.; Liu, Y.; Wang, Y.T. Teaching based on augmented reality for a technical creative design course. *Comput. Educ.* **2015**, *81*, 221–234. [CrossRef]

20.   Fabregas, E.; Farias, G.; Dormido-Canto, S.; Esquembre, F. Developing a remote laboratory for engineering education. *Comput. Educ.* **2011**, *57*, 1686–1697. [CrossRef]

21.   Djouad, T.; Mille, A. Observing and Understanding an On-Line Learning Activity: A Model-Based Approach for Activity Indicator Engineering. *Technol. Knowl. Learn.* **2018**, *23*, 41–64. [CrossRef]

22.   Rio, C.J.; Pastor, M.E.M.; Calle, R.C.; Robaina, N.F. Academic Performance in Higher Education and its Association to Active Participation in the Moodle Platform. *Estudios Sobre Educ.* **2018**, *34*, 177–198. [CrossRef]

23.   Kic, P. The course transport, handling and manipulation machinery in engineering education. In Proceedings of the 13th International Scientific Conference on Engineering for Rural Development, Jelgava, Latvia, 29–30 May 2014; Osadcuks, V., Ed.; Latvia University of Agriculture: Jelgava, Latvia, 2014.

24.   Daroczi, M.; Toth, R.; Molnar, C. The influence of Information Technology on Agricultural Machinery. In Proceedings of the 7th International Scientific Conference on Managerial Trends in the Development of Enterprises in Globalization Era (ICoM), Nitra, Slovakia, 1–2 June 2017; Kosiciarova, I., Kadekova, Z., Eds.; Slovak University of Agriculture: Nitra, Slovak, 2017.

25. Yin, C.Q.; Sun, Q.; Wu, J.; Liu, C.Q.; Gao, J. Development of Electrohydraulic Steering Control System for Tractor Automatic Navigation. *J. Electr. Comput. Eng.* **2018**, *2018*, 5617253. [CrossRef]

26. Lee, D.H.; Kim, Y.J.; Choi, C.H.; Chung, S.O.; Inoue, E.; Okayasu, T. Development of a Parallel Hybrid System for Agricultural Tractors. *J. Fac. Agric. Kyushu Univ.* **2017**, *62*, 137–144.

27. Sim, K.; Lee, H.; Yoon, J.W.; Choi, C.; Hwang, S.H. Effectiveness evaluation of hydro-pneumatic and semi-active cab suspensión for the improvement of ride confort of agricultural tractors. *J. Terramechanics* **2017**, *69*, 23–32. [CrossRef]

28. Castagnetti, D.; Bertacchini, A.; Spaggiari, A.; Lesnjanin, A.; Larcher, L.; Dragoni, E.; Arduini, M. A novel ball joint wear sensor for low-cost structural health monitoring of off-highway vehicles. *Mech. Ind.* **2015**, *16*, 507. [CrossRef]

29. Rencin, L.; Polcar, A. Determination of the tractor engine power in the field conditions. In Proceedings of the 23rd International PhD Students Conference (MendelNet), Brno, Czech Republic, 9–10 November 2016; Polak, O., Cerkal, R., Belcredi, N.B., Horky, P., Vacek, P., Eds.; Mendel Univerzity in Brno: Brno, Czech Republic, 2016.

30. De Farias, M.S.; Schlosser, J.F.; Estrada, J.S.; Frantz, U.G.; Rodriguez, F.A. Evaluation of new agricultural tractors engines by using a portable dynamometer. *Cienc. Rural* **2016**, *46*, 820–824. [CrossRef]

31. Pereira-Marin, C.A.; Perez-Mendez, A.; Marin-Darias, D.; Gonzalez-Cueto, O. ExploMaq, software for energetic and economic evaluating of farming machine. *Rev. Cienc. Técnicas Agropecu.* **2015**, *24*, 72–76.

32. Santos, F.L.; de Queiroz, D.M. Simtrac-an application for simulation of traction efficiency of agricultural tractors with front wheel assist. *Acta Sci.-Technol.* **2016**, *38*, 423–430. [CrossRef]

33. March-Andreu, V.; Lozano-Terrazas, J.L. *Mecanización Agraria. Maquinaria Agrícola y forestal*; Llorens: Valencia, Spain, 2014; p. 318. ISBN 978-84-616-8005-4.

34. Callejon-Ferre, A.J.; Lopez-Martinez, J.A.; Lopez-Perez, A.I. *Manual de Ejercicios y Cuestiones de Clase de la Asignatura de Motores y Máquinas Agrícolas Adaptado al Espacio Europeo de Educación Superior*; Universidad de Almería: Almería, Spain, 2008; pp. 1–190. ISBN 978-84-691-8107-2.

35. Matcho, J. *Using Delphi 2*, Special Edition ed; Prentice Hall: Bergen, NJ, USA, 1996.

36. Evans, M.D.; Clark, R.L.; Manor, G. An improved traction model for ballast selection. *Trans. ASAE* **1991**, *34*, 773–780. [CrossRef]

37. Shibusawa, S.; Sasao, A. Traction data analysis with the traction prediction equation. *J. Terramechanics* **1996**, *33*, 21–28. [CrossRef]

38. Sharma, A.K.; Pandey, K.P. Traction data analysis in reference to a unique zero condition. *J. Terramechanics* **1998**, *35*, 179–188. [CrossRef]

39. Raheman, H.; Jha, S.K. Wheel slip measurement in 2WD tractor. *J. Terramechanics* **2007**, *44*, 89–94. [CrossRef]

40. Huang, Y.J.; Khajepour, A.; Zhu, T.J.; Ding, H.T. A Supervisory Energy-Saving Controller for a Novel Anti-Idling System of Service Vehicles. *IEEE/ASME Trans. Mechatron.* **2017**, *22*, 1037–1046. [CrossRef]

education
sciences

MDPI

Article

A Reverse Engineering Role-Play to Teach Systems Engineering Methods

Alessandro Bertoni

Department of Mechanical Engineering, Blekinge Institute of Technology, 371 41 Karlskrona, Sweden; alessandro.bertoni@bth.se; Tel.: +46-455-385-502

Received: 6 December 2018; Accepted: 28 January 2019; Published: 31 January 2019

Abstract: Students engaged in systems engineering education typically lack experience and understanding of the multidisciplinary complexity of systems engineering projects. Consequently, students struggle to understand the value, rationale, and usefulness of established systems engineering methods, often perceiving them as banal or trivial. The paper presents a learning activity based on a three-stage reverse engineering role-play developed to increase students' awareness of the importance of correctly using systems engineering methods. The activity was developed and integrated in the Systems Engineering course given at Blekinge Institute of Technology. Its effectiveness was analyzed through semistructured self-reflection reports along with two editions of the course. The results showed the development of students' understanding of how to use systems engineering methods. In particular, the students realized the need to deliver detailed and easy-to-read models to the decision makers. This result was in line with the achievement of some of the intended learning outcomes of the course.

Keywords: systems engineering; education; role-play; self-reflection; reverse engineering; active learning; CDIO; learning activity

1. Introduction

The evolution of humanity has always been characterized by a continuous attempt to improve the quality of life and to create better living conditions. The search for new ways of satisfying fundamental needs has often led to the engineering of complex systems that, in order to be successful, needed to be socially acceptable and provide value. At present, the term systems engineering (SE) is commonly used to refer to the engineering effort of satisfying articulated sets of needs from different stakeholders, with the intent of developing solutions providing value to an overarching system [1]. The concept of SE was coined in response to the need to develop increasingly complex systems in the aerospace industry [2]. The application of methods and tools for SE has largely impacted, for instance, the development of aircraft or satellites (see References [3,4]). The increasing socioeconomic challenges related to globalization, population growth, economic interdependence, and sustainability are stressing the need for SE competencies to be applied beyond the aerospace context [1]. While the need for such competences is increasingly recognized in industry and research, those are still poorly addressed from a pedagogical perspective for what concerns the education of new systems engineers. As highlighted by the International Council of Systems Engineering, current SE programs focus on practice, with little emphasis on underlying theory, especially for what concerns human and social sciences [1].

SE projects require multidisciplinary skills and cross-functional design teams, including a wide set of disciplines, such as design, manufacturing, system analysis, knowledge management, and sustainability analysis (see, for instance, References [5–7]). Different SE methods to structure, formalize, and validate knowledge have been developed by both researchers and industrial practitioners since the 1980s. Those have been applied in a variety of industrial contexts to increase standardization and reduce the risk of misinterpretation and ambiguity [8]. However, while SE methods are easily perceived as beneficial in complex industrial contexts, research has shown that their benefits are not evident for students engaged in SE education. This is because students do not commonly have an understanding of the system complexity and perceive the methods as "open doors". Students lack the background to understand what the methods offer and, therefore, risk perceiving them as trivialities [9].

This paper addresses this issue by proposing a learning activity developed to increase students' awareness on the importance of knowledge formalization and communication in early design through established SE methods. The activity, named "Reverse Engineering Role-Play", was inspired by the principle of active learning [10] and was developed in the frame of an educational initiative to promote the CDIO (Conceive—Design—Implement—Operate) framework [11]. The learning activity featured the active participation of students in a role-play in which they acted as development engineers in a reverse engineering analysis of a real product (stage 1), proposed a new design of the product (stage 2), and assumed the role of managers in decision making, evaluating, and choosing the design to be promoted for further development (stage 3). The knowledge created during the three stages was formalized through SE methods, and it was exchanged between different teams during the role-play.

The purpose of this research was to study the effectiveness of the proposed role-play in developing students' critical thinking and self-confidence in choosing and applying systems engineering methods. Data about the effectiveness of the activity in contributing to the intended learning outcomes were collected through semistructured individual self-reflection reports. The results showed an increased students' understanding of the desired quality and level of detail needed for SE methods to be effective in supporting design decision making. Such results were supported by the students' proposal of concrete actions to be taken to improve the quality of their own SE models produced in stage 1.

The structure of the paper is as follows. Section 2 describes the research approach and the educational context, detailing the course structure and defining how the proposed learning activity fits with the intended learning outcomes. Section 3 presents the challenges in SE education and the current effort in reforming the overall engineering education in light of the CDIO framework. Section 4 details the reverse engineering role-play, describing activities, support material, methods used, and how the knowledge was shared. Section 5 reports the results of the data analysis and discusses them in relation to the intended learning outcomes. Section 6 draws the final conclusions.

2. Research Approach and Educational Context

The research presented investigated SE education literature in search for teaching initiatives following the guidelines promoted by the CDIO framework for engineering education. Research data concerning the use of the proposed learning activity were collected during two editions of the Systems Engineering course that took place in the Mechanical Engineering department at Blekinge Institute of Technology" in 2016 and 2017. In total, 46 students enrolled in Master level studies in Mechanical Engineering, Industrial Economics, and Product-Service Systems Innovation were involved. Data about students' activities and performances were analyzed by means of retrospective analysis of individual semistructured self-reflection reports, filled in as the last activity of the role-play. The self-reflection reports were analyzed by categorizing the content into five categories, namely: The reason for a choice, the confidence in the decision, the wish for information, the wish for communication, and the individual learning.

2.1. Description of the Systems Engineering Course

The Systems Engineering course at Blekinge Institute of Technology is designed for university students at Master level. It addresses both SE theories and SE practical skills needed for practical application of SE methods and tools in practice. The course accounts for 7.5 ECTS points in the European Credit Transfer Systems. The European Credit Transfer Systems is based on learning achievements and students' workload, with 60 ECTS points corresponding to a full academic year with a workload range from 1500 to 1800 hours. Every course accounts for a specific number of ECTS points, defining the expected amount of workload for the students, subsequently guiding the definition of the learning objectives and the duration of the course. The intended learning outcomes of the course are listed here:

1. Understand what SE is and what characteristics sets it apart from other development approaches;
2. Assess why and when an SE approach is desirable;
3. Be able to undertake an SE program definition study and define an SE plan for a realistic project;
4. Be able to develop an SE concept design as the basis for further detailed design;
5. Apply SE tools (e.g., requirements development and management, robust design, design structure matrix) to realistic problems;
6. Know how to proactively design for system lifecycle targets.

The course is structured in two internal modules which run in April and May. In the first module, time is spent in classical frontal lectures, where the basic concepts of SE are introduced. As a conclusion to the first module, a class activity is organized in two sessions. This aims to let students experience and test SE methods prior to their extensive application in the second module. In the second module, approximately 90 hours are spent on a course project dealing with an SE problem provided by a partner company. Additional project-related lectures and guest lectures by industrial representatives are eventually run based on the topic of the project. The reverse engineering role-play presented in the paper was run in between the two modules as a class activity to let the students experience the application of SE methods and learn and reflect on their characteristics, quality, and need for standardization. Figure 1 visualizes the high-level structure of the course.

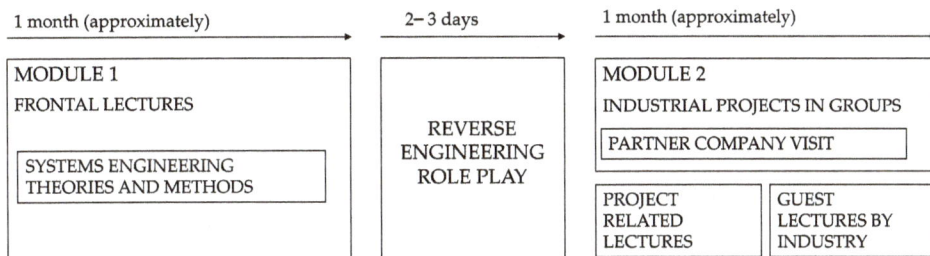

Figure 1. Modules of the Systems Engineering course.

3. Framing the Proposed Approach into the Systems Engineering Education Challenges and the CDIO Initiative

The academic discussion about the role of SE education goes back to the 1980s. Most of the early contributions focused on specific contents to grant an education that was in line with the professional needs of industries interested in employing the newly graduated students [12]. Asbjornsen and Hamann [12], in their work toward the definition of a unified SE education, listed a few requirements in relation to the SE process outlined by Asbjornsen [13]. According to those requirements, future systems engineers would need to understand and analyze a broad and multidisciplinary knowledge-base and, at the same time, would need the skills to work with deep quantitative knowledge in specialized areas. The skills of a systems engineer were also defined as strongly related to the ability to understand other disciplines and to explain one's own. This is related to the ability to learn, share, and communicate new knowledge, ultimately lifting the need for systems engineers to be able to understand and analyze the human interrelations inside an organization (as also partially described by Senge [14] concerning practices in learning organizations). Asbjornsen and Hamann [12] also highlighted the need for "softer" skills, such as loyalty, individual responsibility, and global and environmental concerns. Their work, however, did not prescribe specific learning activities to address such multidisciplinary, neither indicated effective ways to teach SE to university students. A more recent work by Muller and Bonnema [9] focused on finding a teaching method fitting the limited students' experience in the SE context. The authors investigated how to effectively teach SE and listed the differences between teaching SE compared to the other engineering disciplines. Specifically, five characteristics were described as critical:

- The broader scope, encompassing technical problems and many nontechnical issues at the same time;
- The ill-defined nature of the problems;
- The presence of unknowns, uncertainties, contradictions, and ambiguities in the problem definition;
- The presence of problems and solutions that are so heterogeneous that methods, techniques, and formalisms often have to be adapted to the situation;
- The absence of a single unique "best" answer, given the many-dimensional field of stakeholders and concerns.

The effectiveness of teaching engineering disciplines is a topic that goes beyond the focus of SE and concerns both the structure of the learning activities in the class and the learning styles of the students. About the latter, different models have been proposed in the literature (e.g., Myers–Briggs type indicator [15], Kolb learning style model [16], Felder–Silverman learning style [17]). Felder [18] listed a set of recommendations for engineering education to address the larger possible set of students learning styles, underlying, among others, the benefits of the following approaches:

- Teach theoretical material by first presenting phenomena and problem that related to the theory;
- Balance conceptual information with concrete information;
- Use physical analogies and demonstrations;
- Give occasionally experimental observations before presenting general principles;
- Provide class time to think about the material presented.

A parallel stream of literature focused on the analysis of the effect of active learning initiatives on students' motivation and engagement in their own learning. In the early 1990s, a book by Bonwell and Eison examined the major contributions on active learning, concluding that its application improved students' attitude as well as their thinking and writing ability [10]. Further, the review by Prince [19] highlighted the extensive empirical support that active learning has collected during the years and showed the induced benefits concerning the students' improved capability to remember the topic of a lecture.

The CDIO international educational framework defines engineering education as a combination of conceive, design, implement, and operate activities [11]. Learning activities combining active learning, direct participation of students in concrete design tasks, and the participation to workshops and project-oriented group tasks constitute the backbone of the CDIO philosophy. Since its formulation, more than 150 academic institutions have joined the CDIO initiative [20] with the vision that "engineering graduates should be able to: Conceive—Design—Implement—Operate complex value-added engineering systems in a modern team-based engineering environment to create systems and products" [21].

The role-play is a form of active learning where students work through a given scenario adopting different personas and interacting in their assumed role [22]. Literature shows role-play to be an approach that is particularly effective for learning about complex systems and for maintaining students' engagement while avoiding anxiety and increasing independent learning [22]. Research studies interviewing students several months after the closure of a course show that students tend to remember more information from the role-plays than from other lectures [23]. The role-play was also described as a valuable approach to enhance knowledge acquisition, particularly when it includes the observation and the acquisition of information from others [24]. The potential of role-plays in higher education was also explored by Rao and Stupans [25], proposing a typology of role-play learning designs enabling teachers to select the role-play approach that would best fit the intended learning outcomes.

The reverse engineering role-play presented in this paper is an instantiation of the effort to reform the learning activities of the SE course in the frame of the CDIO initiative. It uses active learning and self-reflection reports to increase students' engagement and understanding of the topic and, at the same time, integrates Felder's recommendations on addressing different students' learning styles.

4. Description of the Learning Activity

The reverse engineering role-play was designed to address the intended learning outcomes number three and five of the course (see Section 2.1). In particular, the activity focused on providing a realistic project in which students can apply SE methods and subsequently reflect on their benefits, drawbacks, and challenges. The proposed learning activity aimed to foster students' understanding of the importance of effective information communication by means of established methods in SE decision making. To achieve this goal, the role-play was organized as a two-day activity in which the students acted initially as members of an engineering team dealing with the reverse engineering of a product (stage 1), assuming later (stage 2) the role of members of a product innovation team in charge of the redesign of the available product. The two stages were followed by an individual activity, in which each student was asked to take the role of a manager to select, among the set of proposed innovations, the most relevant design to be further developed. This was done to replicate a cross-functional development scenario in which designers and decision makers are two different groups of stakeholders.

The role-play took place one month after the beginning of the course, when the students had already acquired knowledge about SE methods and tools (e.g., quality functional deployment, functional analysis, functional decomposition, requirement definition, N^2 diagrams) and had previously seen examples of reverse engineering activities. The following subsections describe in detail each stage of the reverse-engineering role-play.

4.1. Stage 1: Reverse Engineering and Knowledge Formalization

The activities of stage 1 were introduced with a 15-min presentation describing the reference products. The students were randomly divided into teams of 3 (and occasionally 4) members. Two products were proposed as subjects of reverse engineering, namely an electrically-powered coffee machine and an electrically-powered orange juicer (Figure 2). Each team received either the coffee machine or the orange juicer, the necessary instruments to disassemble and test the product, and the original packaging used for transportation. The choice of rather "simple" products, such as a coffee machine and an orange juicer, was driven by the necessity to give the students a product they could easily relate to and that was possible to analyze in a restrict timeframe. In total, 15 student teams took part in stage 1, six of which worked on the reverse engineering of the coffee machine and nine worked on the reverse engineering of the orange juicer.

Figure 2. Picture of the two products to be reverse engineered: A coffee machine (**left**) and an orange juicer (**right**).

The reverse engineering kicked off with the teams receiving the physical product, with the assignment to deliver its functional decomposition and its functional analysis, including the functional flow block diagram and an N^2 diagram [3]. Those models needed to be delivered following the shared SE standards (previously taught in the course lectures), and they needed to be uploaded on a common databased created for the activity. The activity started at 10:00 a.m., and no specific time limit was assigned. However, the teams were asked to gather again at 11:45 a.m. to verify the status of advancement of the work and were requested to upload the results on the database before the end of the day. Figure 3 illustrates the activities and the material produced in stage 1.

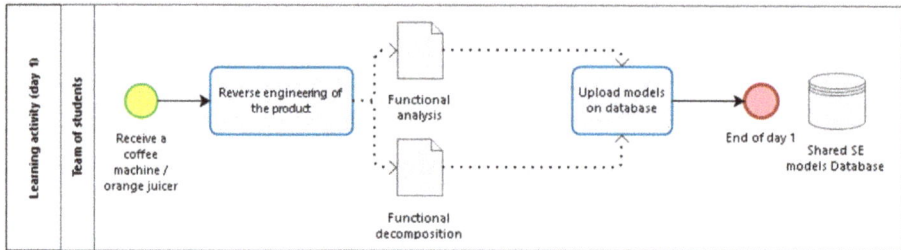

Figure 3. Activities and material produced in stage 1.

4.2. Stage 2: Propose Product Innovations

Stage 2 featured the same time setting and the same team composition of stage 1. It started with a brief introduction of the assignment of the day given by the teacher. Stage 2 took place on a different day than stage 1; this was decided to avoid stressing the delivery of the models during stage 1. In stage 2, the teams changed their role, not dealing with reverse engineering any longer but rather focusing on proposing product innovations. The peculiarity of stage 2 stemmed from teams switching the reference product of their work, that is, students' teams that performed the reverse engineering of the coffee machine were asked to propose a new design of the orange juicer, while teams previously working on the orange juicer were asked to propose a new design of the coffee machine. To fill the knowledge gap given by the product switch, the teams were given access to the database with all the models produced during stage 1 by their classmates. This setting was proposed to replicate a typical situation in SE projects, in which different design teams take care of different stages of the development and need to rely on information that is produced and shared by other teams.

After accessing the database with SE models, the teams received a list of needs and problems about the coffee machine and the orange juicer. This was artificially developed by the teacher for the purpose of the activity and encompassed a broad set of stakeholders, including customers, logistics, production, and marketing. The list of needs and problems for both products is available in Appendices A and B, showing, in the first column, which stakeholder expressed a specific need and, in the second column, the formulation of the need or problem. In total, eight statements for the coffee machine and twelve for the orange juicer were provided. No further categorization or hierarchical analysis of the statements were provided to the students. The teams were asked to produce and document a new design based on the knowledge base. The requirements for the activity were to deliver a sketch of the new product, with related textual explanation, and to fill in a self-assessment report template (available in Appendix C). The structure of the self-assessment report was similar to the previous list of needs and problems, although it added three columns, in which the teams were asked to evaluate: How much the new design fulfilled the provided needs in relation to the actual product (column 3), how confident the team was in the assessment for each need/problem (column 4), and how experienced the designers were in using the coffee machine or the orange juicer (column 5). The report was complemented by a visual indication on how to self-assess the new design based on a scale from 1 to 9, using the original product as a baseline. Furthermore, a table mimicking the knowledge maturity scale proposed in SE literature by Johansson et al. [26] was provided to guide the assessment of the confidence of the teams in column

4. After completing the tasks, all the documentation was uploaded to the shared database to enable the evaluation of the decision maker. Figure 4 illustrates the activities and the material produced in day 2.

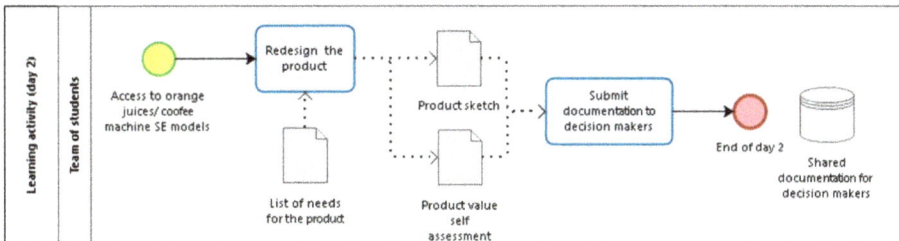

Figure 4. Activities and material produced in stage 2.

4.3. Stage 3: Decision Making and Self-Reflection

In the last part of the role-play, the students were asked to act as individual decision makers going through the available documentation and selecting the most promising design of coffee maker and orange juicer to be further developed. The students were asked to individually select one design for each product and justify their choice in written text. Furthermore, the students were also asked to self-reflect on the models that the teams produced during the activity, discussing if and how those were useful in the decision-making activity. The following questions where provided as a guideline for the students' reflections:

- Was the information enough to be confident with your choice? If not, what kind of information would you have asked the design teams? And, eventually, in which form would you have such information communicated to you?
- Have you found one or more functional flow block diagrams more useful than others? If yes, why? If no, what would you expect to learn more from such diagrams? If you would do your diagram again, would you do it differently?
- Have you found one or more N^2 matrices more useful than others? If yes, why? If no, what would you expect to learn more from such diagrams? If you would do your N^2 matrix again, would you do it differently?
- Have you found one or more functional decompositions more useful than others? If yes, why? If no, what would you expect to learn more from such diagrams? If you would do your functional decomposition again, would you do it differently?

The reflections were formalized in a text document and submitted to the teacher as part of the course examination valuable for the final grade.

5. Data Analysis and Results

The documentation generated from stage 3 constituted the set of raw data that was analyzed to verify the effectiveness of the proposed learning activity. The reports were analyzed looking for answers to five questions, namely:

1. What were the main reasons to choose a design?
2. Was the information enough to be confident in your decision?
3. What kind of information would you have asked the design teams?
4. In which form would you have preferred to have such information communicated to you?
5. If you would do your models again, what would you do differently?

Students were free to provide open answers, indicating one or more aspects that they would have liked to improve or that influenced their decision; thus, the analysis of the data accounted for possible multiple answers. The following subsections describe the findings for each question.

5.1. Reason to Choose a Design and Confidence in the Choice

All 46 students provided the reasons to choose a particular design. Figure 5 shows the distribution of answers to question 1.

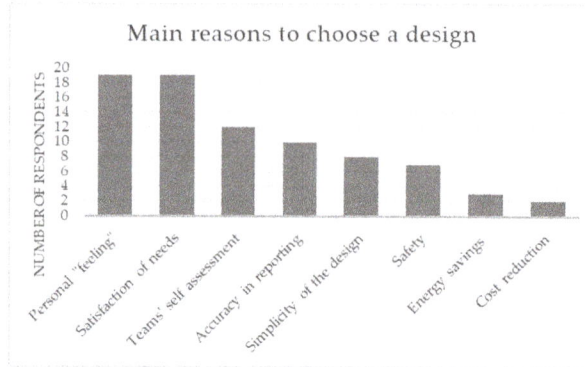

Figure 5. Distribution of the students' feedback on the main criteria to choose a design.

The data show two dominant reasons to select a proposed design, namely the "personal feeling" of the decision maker and the perception that the new design is satisfying the needs of different stakeholders. Even if dominant in relation to the other response, those two reasons were mentioned only by 19 of the 46 students (corresponding to the 41% of the sample). Despite the satisfaction of the needs being central, only 12 of the 46 students admitted having considered the teams' self-assessment report concerning needs satisfaction in their choice, while 10 out of 46 students were influenced in their choice by the accuracy of the report submitted. Such data are interesting to be analyzed in relation to the students' confidence in making decisions. From the 40 students that explicitly discussed their confidence, 27 (corresponding to the 67.5%) stated that there was not enough information to be confident in their decision, although they were required to make it anyway. These findings are particularly relevant in the frame of the different roles that the students took in the role-play. This highlights the challenges that decision makers in an SE context face when asked to make decisions with product and systems information poorly defined and formalized. In other words, 27 out of 40 students realized that, to make a confident decision, the SE models produced by the class needed to have a higher level of detail and precision. Such data suggested the activity was effective in allowing students to experience the typical dynamics of decision making in an SE context.

5.2. Type of Information to Be Further Asked to the Design Team

The students were further asked to state what kind of additional information they would have asked the design team to increase their confidence in the decision. Forty-one students addressed this question in their report, and the distribution of answers is shown in Figure 6. Out of 41 students, 20 recognized the need for enhanced visual representation of the designs. Sixteen students (i.e., 39% of the respondents) also highlighted the need for decision makers to receive more accurate information about the product parts, and 10 students expressed the desire to have a description of the product working process. Eight students requested an explanation of the rationale for the design choices, and seven would have preferred a more detailed analysis of the stakeholders' needs. Of particular interest in relation to the intended learning outcomes is the students' awareness of the necessity of having more accurate information about the product parts and the product working process. Those two aspects are respectively addressed by the functional decomposition and the functional analysis, models previously experienced and produced by the students in stage 1. These data show that, despite experiencing the models in stage 1, many students failed to deliver appropriate SE models during stage 2. The recognition of such failure happened in stage 3, during the decision-making phase. From a pedagogical perspective, such a self-recognition process should be regarded as a positive result of the learning activity, in line with the desired intended learning outcomes.

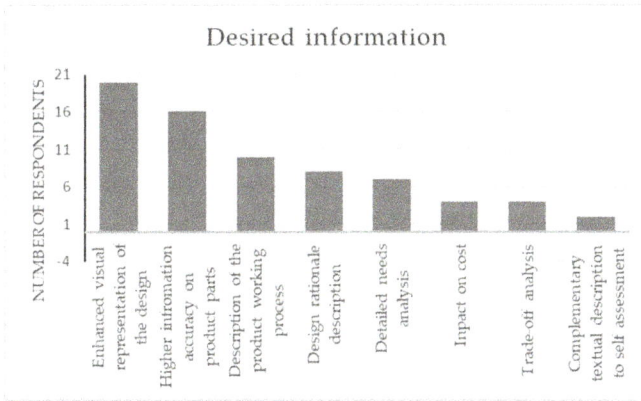

Figure 6. Type of information further desired from the design teams.

Out of the 41 students addressing the need for more information, 28 detailed in which specific form they would have liked to receive it. No specific form emerged as dominant in the answers, with 9 students expressing the desire for an oral presentation, 7 wishing to access CAD models and physical prototypes, and few other students expressing the desire for more concise description, pictures, and diagrams. The absence of dominant results does not allow to draw any specific pedagogical reflection about this aspect.

5.3. What Would You Do Differently

The last question was the most relevant for what concerned the identification of the learning results from the reverse engineering role-play. In light of their experience as innovators in stage 2 and decision makers in stage 3, the students were asked to reflect on what they would do differently if they came back to stage 1 and redid the functional models and the functional decomposition again. The question was of primary importance to understand if the learning activity contributed to the objective of raising students' awareness about the importance of using defined and standardized SE methods for knowledge communication in cross-functional teams. Figure 7 shows the distribution of the answers given by the 32 students that explicitly addressed this question in their reports.

The majority of the students (24 over 32, equal to 75%) stated that they would make a bigger effort in adding details to the SE models produced in stage 1. Eleven students (35% of the total) also raised the necessity for the new models to be more readable than the previous ones, also adding pictures to facilitate the communication (8 students). In light of such data, it is possible to state that the students developed a critical perspective toward their own work, reflecting on the necessity to use SE methods in a way that allows easier knowledge and information communication at different functional levels and for different uses. This is shown by the shared understanding that the level of detail of the models that the teams delivered during stage 1 did not reflect the level of detail that was later needed by the other teams when acting as product innovators and decision makers. Furthermore, the recognition of the need for more readable SE models can reflect the understanding of the multidisciplinary application of such methods in an SE context. From an overall perspective, the finding from the data analysis described a shared understanding about the necessity to use well-formalized, detailed, and effectively communicated methods during SE, suggesting that the learning activity was effective in avoiding the risk of the SE method being considered as a triviality by students lacking contextual SE knowledge (a concern also expressed by Muller and Bonnema [9]).

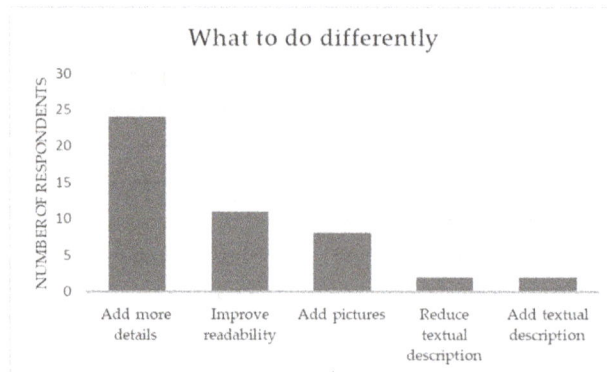

Figure 7. Distribution of answers for "What I would do differently if I would do the models again".

6. Concluding Remarks

The evolving needs of society and industry have constantly changed the way in which engineering education is performed. Engineering education continuously strives to educate students about the skills and abilities needed to efficiently and effectively create value in society. Among the different areas of improvement, this paper has focused on a specific aspect, that is, the capability of engineers to work in an SE context.

This paper presented a learning activity for SE education that addresses parts of the challenges raised in the educational literature described in Section 3. The paper focused on the necessity of conveying the relevance of SE methods to students that struggle to perceive the usefulness of such models. The analysis of the data collected from the students' self-reflections showed an increased awareness of the students on the necessity of extensively reviewing their initial use of SE methods. The addition of details and the improvement of readability emerged as key actions to provide better-formulated information to decision makers, ultimately enabling more confident design decisions. This is interpreted as a positive result in relation to the satisfaction of the course-intended learning outcomes. Concerning the limitation of the research, it must be highlighted that the data presented were collected from a sample of 46 students, and this can limit the generalizability of the findings. Engineering literature describes several examples of experiments with university students featuring an even smaller number of participants (see References [27,28]), although the verification of the effectiveness of the described educational approach would benefit from a larger data sample, also including students from different institutions and cultures.

The reverse engineering activity was based on commercial products not typically developed through an SE process. This allowed the students to model the behavior of a product in a usage context that was familiar to them. The benefit of this was limiting the risk of including too many assumptions about the functional decomposition and the functional analysis of the products. The adoption of the new learning activity in the course was framed in the university effort to move toward the transformation objectives highlighted by the CDIO framework. In line with the active learning principle, the role-play was designed around an important learning outcome of the course (i.e., the capability of applying SE tools to realistic problems) and was designed to promote the engagement of the students in the course. In conclusion, the proposed learning activity was shown to promote students' critical thinking and to increase students' self-confidence and understanding of the usefulness of SE methods, providing the students with an answer to the question: "Why should I care about learning this?"

Future evolution of the proposed learning activity concerns the exploration of the possibility to run a larger-scale reverse engineering role-play involving SE students of multiple academic institutions, in order to replicate the challenges given by distributed and globally located design teams.

Funding: This research received no external funding.

Conflicts of Interest: The author declares no conflict of interest.

Appendix A

List of needs to consider for the redesign of the coffee machine as distributed to the students.

Stakeholder	Needs or problems
Customer	The product does not give a sense of stability when used.
	If you forget it on, it will consume a very high amount of electricity.
	Some customers do not like that the coffee flows along a metal spring before reaching the jar both for hygiene (the spring can rust) and coffee taste.
	The short cord creates the need to have the machine positioned close to an electrical outlet.
	The handle and the lid of the jar are reported to be difficult to clean.
Logistics	The logistic is operated by "ICA Logistics" and would benefit by a reduction in dimension of the product.
Production	Reducing the number of components would make the product cheaper to produce.
	A (marginal) increase in production cost (material/features) is acceptable if this reduces the production time.
Marketing	The coffee machine will be marketed as "the machine for the family", and the marketing needs one or more design features to claim it to be "good/safe for children".

Appendix B

List of needs to consider for the redesign of the orange juicer as distributed to the students.

Stakeholder	Needs or problems
Customer	Some customers complain about the middle container to be small.
	Some customers complain about the citrus holder on the handle to be difficult to clean.
	The pulp of the oranges gets stuck between the filter and the head cone.
	Some customers say they will not buy the product because too big to be placed in their small kitchen.
	The short cord creates the need to have the squeezer positioned close to an electrical outlet.
	Big oranges are not well squeezed because the lateral part does not get in contact with the cone.
	The small circular seal of the middle container feels cheap and it is difficult to clean.
Logistics	Reducing dimensions and weight would save storage and transport cost.
Production	The use of non standard screws slows down the production process and reduce commonality with other products.
	The presence of a double cone in the head (there is a smaller one hided in the bigger one) raises production costs and requires more strict tolerances.
Marketing	There is no budget for advertising, the product needs to attract customers' attention in the shop.
	Marketing says the product "lacks of personality", that it is "boring".

Appendix C

Stakeholder	Needs or problems	Self assessment	How confident are you in the assessment? (1 to 5)	Personal experience of coffe machines
Customer	The product does not give a sense of stability when used.			Person #1
	If you forget it on it will consume a very high amount of electricity.			
	Some customer do not like that the coffee flow along a metal spring before reaching the jar both for hygiene (the spring can rust) and cofee taste.			
	The short cord creates the need to have the squeezer positionet close to an electrical outlet			Person #2
	The handle and the lid of the jar are reported to be diffult to clean			
Logistics	The logistic is operated via ICA logistic and would benefit by a reduction in volume			Person #3
Production	Reducing the number of component would make the product cheaper to produce			
	A (marginal) increase in production cost (matrial/features) is acceptable if this reduces the production time.			Person #4
Marketing	The coffee machien will be marketed as "the machine for the family", and the marketing needs one or more design feature to claim it to be "good/safe for children"			

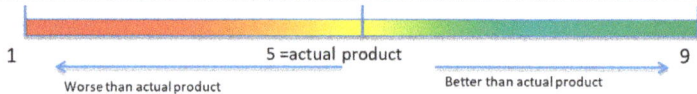

1   5 =actual product   9

Worse than actual product   Better than actual product

1	2	3	4	5
INFERIOR Risk of not correct assessment	in between 1 and 3	**INTERMEDIATE** The assessment shall be trustable but it is difficult to verify	in between 3 and 5	**HIGH** The assessment is reliable and can be verified

Template for the self-assessment report as distributed to the students.

References

1. INCOSE. *A World in Motion: Systems Engineering Vision 2025*; International Council on Systems Engineering: San Diego, CA, USA, 2014.
2. Blanchard, B.S.; Fabrycky, W.J.; Fabrycky, W.J. *Systems Engineering and Analysis*; Prentice Hall: Englewood Cliffs, NJ, USA, 1990; Volume 4, ISBN 978-0132217354.
3. Shishko, R.; Aster, R. *NASA Systems Engineering Handbook*; NASA Special Publication 6105; The National Aeronautics and Space Administration (NASA): Washington, DC, USA, 1995; ISBN 978-1680920895.
4. Estefan, J.A. Survey of model-based systems engineering (MBSE) methodologies. *INCOSE MBSE Focus Group* **2007**, *25*, 1–12.
5. Isaksson, O.; Kossmann, M.; Bertoni, M.; Eres, H.; Monceaux, A.; Bertoni, A.; Wiseall, S.; Zhang, X. Value-Driven Design–A methodology to link expectation to technical requirements in the extended enterprise. In Proceedings of the INCOSE International Symposium, Philadelphia, PA, USA, 24–27 June 2013; Volume 23, pp. 803–819. [CrossRef]
6. Monceaux, A.; Kossmann, M.; Wiseall, S.; Bertoni, M.; Isaksson, O.; Eres, H.; Bertoni, A.; Rianantsoa, N. Overview of value-driven design research: Methods, applications, and relevance for conceptual design. *Insight* **2014**, *17*, 37–39. [CrossRef]
7. Bertoni, A.; Dasari, S.K.; Hallstedt, S.; Andersson, P. Model-based decision support for value and sustainability assessment: Applying machine learning in aerospace product development. In Proceedings of the 15 International Design Conference, Dubrovnik, Croatia, 21–24 May 2018; pp. 2585–2596. [CrossRef]
8. INCOSE. *Systems Engineering Handbook: A Guide for System Life Cycle Processes and Activities*; John Wiley and Sons, Inc.: Hoboken, NJ, USA, 2015; ISBN 978-1-118-99940-0.
9. Muller, G.; Bonnema, G.M. Teaching systems engineering to undergraduates; Experiences and considerations. In Proceedings of the INCOSE International Symposium, Philadelphia, PA, USA, 24–27 June 2013; Volume 23, pp. 98–111. [CrossRef]

10. Bonwell, C.C.; Eison, J.A. *Active Learning: Creating Excitement in the Classroom*; ASHEERIC Higher Education Report No.1; George Washington University: Washington, DC, USA, 1991; ISBN 978-1878380081.
11. Berggren, K.F.; Brodeur, D.; Crawley, E.F.; Ingemarsson, I.; Litant, W.T.; Malmqvist, J.; Östlund, S. CDIO: An international initiative for reforming engineering education. *WTE&TE* **2003**, *2*, 49–52.
12. Asbjornsen, O.A.; Hamann, R.J. Toward a unified systems engineering education. *IEEE Trans. Syst. Man Cybern. C* **2000**, *30*, 175–182. [CrossRef]
13. Asbjornsen, O.A. Technical Management-a Major Challenge in Industrial Competition. *Chem. Eng. Progr.* **1988**, *84*, 27–32.
14. Senge, P.M. *The Fifth Discipline: The Art and Practice of the Learning Organization*; Doubleday: New York, NY, USA, 1990; ISBN 0-385-51725-4.
15. Lawrence, G.D. *People Types & Tiger Stripes*; Center for Applications of Psychological Type, Inc.: Gainesville, FL, USA, 1993.
16. Kolb, D.A. *Experiential Learning: Experience as the Source of Learning and Development*; FT Press: Upper Sadle River, NJ, USA, 2014; ISBN 978-0133892406.
17. Felder, R.M.; Silverman, L.K. Learning and teaching styles in engineering education. *Eng. Educ.* **1988**, *78*, 674–681.
18. Felder, R.M. Matters of style. *ASEE Prism USA* **1996**, *6*, 18–23.
19. Prince, M. Does active learning work? A review of the research. *J. Eng. Educ.* **2004**, *93*, 223–231. [CrossRef]
20. CDIO Members Schools. Available online: http://www.cdio.org/cdio-collaborators/school-profiles (accessed on 5 December 2018).
21. CDIO Vision. Available online: http://www.cdio.org/cdio-vision (accessed on 5 December 2018).
22. Russell, C.; Shepherd, J. Online role-play environments for higher education. *Brit. J. Educ. Technol.* **2010**, *41*, 992–1002. [CrossRef]
23. DeNeve, K.M.; Heppner, M.J. Role play simulations: The assessment of an active learning technique and comparisons with traditional lectures. *Innovat. High. Educ.* **1997**, *21*, 231–246. [CrossRef]
24. Van Ments, M. *The Effective Use of Role-play: Practical Techniques for Improving Learning*; Kogan Page Publishers: London, UK, 1999.
25. Rao, D.; Stupans, I. Exploring the potential of role play in higher education: development of a typology and teacher guidelines. *Innov. Educ. Teach. Int.* **2012**, *49*, 427–436. [CrossRef]
26. Johansson, C.; Hicks, B.; Larsson, A.C.; Bertoni, M. Knowledge maturity as a means to support decision making during product-service systems development projects in the aerospace sector. *Proj. Manag. J.* **2011**, *42*, 32–50. [CrossRef]
27. Cash, P.; Elias, E.; Dekoninck, E.; Culley, S. Methodological insights from a rigorous small scale design experiment. *Des. Stud.* **2012**, *33*, 208–235. [CrossRef]
28. Bertoni, A. Analyzing Product-Service Systems conceptual design: The effect of color-coded 3D representation. *Des. Stud.* **2013**, *34*, 763–793. [CrossRef]

education sciences

MDPI

Article

Experimental Equipment to Develop Teaching of the Concept Viscosity

Modesto Pérez-Sánchez [1,*], Ruzan Galstyan-Sargsyan [2], M. Isabel Pérez-Sánchez [3] and P. Amparo López-Jiménez [1]

[1] Hydraulic and Environmental Engineering Department, Universitat Politècnica de València, 46022 Valencia, Spain; palopez@upv.es
[2] Applied Linguistics Department, Universitat Politècnica de Valencia, 03801 Alcoy, Spain; rugalsar@upvnet.upv.es
[3] Junta de Comunidades de Castilla La Mancha, 45001 Toledo, Spain; miperezs@edu.jccm.es
* Correspondence: mopesan1@upv.es; Tel.: +34-96-387-700

Received: 11 September 2018; Accepted: 19 October 2018; Published: 20 October 2018

Abstract: Some of the subjects have complex concepts, which are currently taught using deductive methods in the first years of University Degree. However, the experience shows the results obtained from students' learning goals were quite low. Therefore, the use of inductive method is a crucial factor to improve students' learning results and re-thinking the way to teach in basic subject of Engineering Bachelor Degree. One example is the subject called Fluid Mechanics, which is present in many Bachelor Degrees. This matter has abstract concepts, which are normally taught by traditional methods. This type of teaching makes difficult to be understood by the student. This research proposes an inductive methodology to work the viscosity concept using an activity. In this test, the student has to carry out some measurements with different fluids using a simple measurement device while they participated actively in the learning.

Keywords: inductive methods; re-thinking the teaching; viscometer

1. Introduction

If the learning process is analysed inside a University environment, particularly in Engineering Bachelor Degree, there are some problems. One of the main problems is the complexity of numerous concepts in some subjects, which are taught in the first courses such as Mathematics, Physics or Fluid Mechanics [1,2]. Currently, this complexity is bigger in the subjects, which use traditional methods. In these methods, the theoretical concepts are taught by the professor and the student has to assimilate it (e.g., master class, lectures, and exercises). This learning causes the students are neither receptive nor motivated to study. Therefore, they do not reach the learning objectives, which are considered by the teaching guide of the subject. Consequently, the use of the active learning methods in which the students participle actively in the development of the learning is recommendable. These methods are based on experimentation (e.g., playing learning, project-based learning, role activities) and they are a possible solution to improve the students' learning. Active learning, collaborative learning, cooperative learning and problem-based learning are active methodologies that use traditional activities partially, but the students learn participating in the learning process actively. A deep review and discussion of these methods was considered by [3]. Therefore, they should be applied to facilitate the understanding of the student at different cognitive levels (lower order thinking skills (*LOTS*) and higher order thinking skills (*HOTS*)). Viscosity, which is taught in Fluid Mechanics is one of those theoretical concepts.

The object of this research is to show a simple activity in which different experiment measurements are developed by students, improving the understanding of the concept of fluid viscosity. The knowledge is acquired by students through activities methodologies, particularly using experiential learning and combining with mathematical equations. The consideration of experimental and the use of mathematics theory allow students to increase the acquisition of competences [4,5]. Besides, they can propose physical models to explain the different viscous nature of various fluids. Therefore, this practice could be implemented in the subject of the Fluids Mechanics in any Bachelor Degree where the matter is taught. If this activity is developed, the students would be given better opportunities to learn the viscosity concept and to analyse its properties using a simple apparatus, which is described. This apparatus could be reproduced in any lab and it can operate with common materials such as water, ketchup or cornstarch. Besides, the research shows the possibility to evaluate the transversal competence 'CT-13.-Specific instrumentation', which is included in the project 'Plan Estrátegico 2015–2020'. This plan is currently being developed by Universitat Politècnica València (UPV). The UPV plan tries to implement these competences inside of the student's curriculum since the transversal competences increase the significance in the pupil's training joined to specific competences. This project tries to evaluate the learning objective using a descriptor replacing traditional methods [2]. This competence develops student's thinking skills and the way he/she behaves trying to solve or analyse a case study.

The aim of this research is to develop an instrument, which can be used by students to assimilate the viscosity concept, combining the experimental learning and the use of the theoretical equations, which are commonly teaching by traditional methods. The objective is to analyse the fluid nature that is, the measurement of the viscosity value is not goal of this research. To measure it, there are many instruments that were previously enumerated and they obtain values more exact. The methodology was applied in a group of students who were studying Engineering Bachelor Degrees related to the Fluid Mechanics topics. Therefore, re-thinking the way to teach better the concept of viscosity and its influence on the fluid behaviour is the objective of the authors. It was the origin to propose the development of a new viscometer. The use of this device was suggested as an activity to help students visualize the viscosity phenomenon and understand the viscosity variations as a function of the fluid nature as well as temperature.

The manuscript has four different sections. Firstly, the introduction describes the research topic. The second section contains the description of the materials and methods. The third section of this paper shows the experimental results obtained by the students. Finally, the manuscript enumerates the main conclusions and applications of this research.

2. The Viscosity in the Engineering Curricula. Re-Thinking as the Professor Teaches this Crucial Concept in Fluid Mechanics

2.1. The Viscosity. How Is It Taught? Looking for Alternatives to Improve the Student´S Learning Process

2.1.1. The Concept of the Viscosity. What Is It? What Is Its Equation That Defines the Fluid Behaviour?

The viscosity is a fluid property, which leads to the existence of shear forces within the fluid and to resistance when the fluid is in contact with a solid boundary. According to Newton [3], the viscosity is directly proportional (Equation (1)) between applied shear and velocities gradient in a fluid. The fluid is a substance, which experiences a continual deformation when a shear is applied over it [6]. The viscosity effect can be observed through in laminar Couette flow. In this experiment, a rectangular portion of a fluid (i.e., the fluid is confined between two sheets) is subjected to shear using a surface in which a parallel force is applied upon. This force is aligned with the free surface direction. When this force is applied, a velocity gradient is generated in the fluid (Figure 1). This force generated a uniform gradient in which the fluid velocity varies between zero and u. The value zero is for the particles, which are in contact with the lower sheet. The value u is for the particles in contact with the upper sheet. The velocity in the sheet is equal to fluid particle's velocity in contact with it due to adhesion condition. This fluid property is defined by the Equation (1), according to Newton's assumption.

$$\tau = \mu \frac{du}{dy} \tag{1}$$

where τ is the shear; μ is the viscosity, and du/dy is the wall-normal gradient of the wall-parallel velocity.

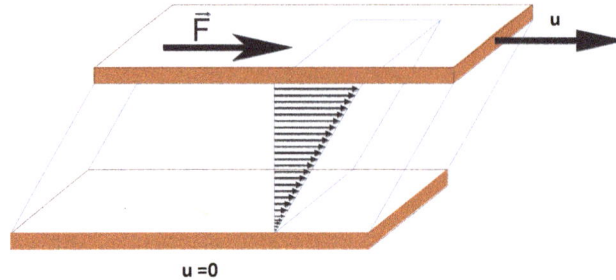

Figure 1. Viscosity defined through Newton's assumption (Adapted from [3]).

2.1.2. The Viscosity. How Is It Taught? Looking for Alternatives to Improve the Student's Learning Process

The Viscosity and Its Traditional Learning

When the viscosity concept is described by using traditional methods, the lesson is generally divided into three parts: concept, type of fluids, and apparatus to measure the property. These concepts are taught using a master class. This type of teaching is a lesson where someone who is an expert at something advises a group of students, using deductive methods. In this type of teaching, the professor teaches concepts, principles, theorems and/or equations during a limited time (normally two or three hours) while the student has to extract the different conclusions and learn the concepts taught by the professor. Once, the theory is taught, the students carry out experimental practices. Along these sessions, the students can check the theoretical concept that were developed in class using commercial viscometers. This type of learning is called cognitive learning and it is not recommended when the student has not yet assimilated the concepts, principles or equations.

The knowledge of the concepts and properties of the viscosity using traditional methods are focused on the following learning objectives. These learning goals are crucial for students' learning in the first unit of the subject of Fluid Mechanics.

The main learning objectives were:

- To understand the concept of the viscosity and the equation that defines it;
- To analyze the viscosity variation according to fluid nature;
- To analyze the viscosity variation according to temperature.

The viscosity is a fluid property, which is similar to friction between solids surfaces. The fluid characteristic mainly depends on: the molecular weight of solute; pressure; suspended matter; and temperature [7]. The cohesion level of the fluid varies according to: (i) physical nature (i.e., liquid or gas), (ii) the relationship between viscosity and concentration of solute is usually direct when the temperature is constant, increasing with the concentration. Otherwise, the viscosity is practically invariable when the pressure varies and the fluid property is inversely proportional to temperature changes.

The viscosity establishes an important classification of the fluids according to their nature. If the viscosity keeps its value constant when the velocity gradient varies, the fluid is defined as Newtonian fluid (e.g., water, oils or glycerine). Otherwise, if the viscosity changes when the velocity gradient is modified, the fluid is denominated as non-Newtonian fluid. This can be classified as: (i) pseudoplastic or Bingham (shear-thinning fluids, or even ketchup); or (ii) dilatant fluid (shear-thickening fluids which increase viscosity at higher rates; e.g., the uncooked paste with water and cornstarch). Figure 2 shows shear stress as a function of strain rate (i.e., Newtonian, pseudoplastic, or dilatant flow). The types of non-Newtonian fluids are completed with the plastic (also called Bingham) and pseudoplastic fluid with yield point.

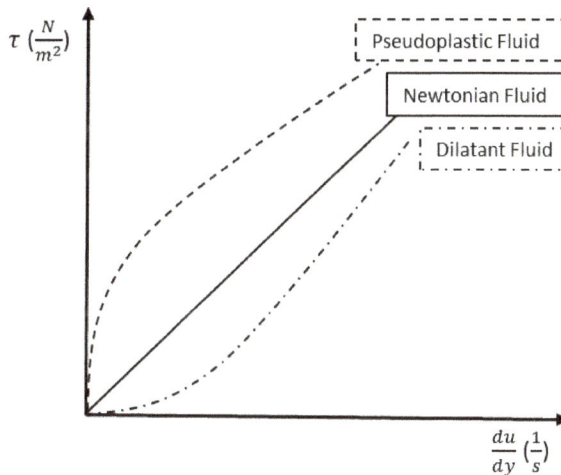

Figure 2. Scheme of the Newtonian and non-Newtonian fluid.

Equation (1) is a particular case of the power law. Independently of nature fluid, the power law defines the viscosity of the fluid, being established by the Equation (2) [8]:

$$\tau = K(\frac{du}{dy})^{n-1} \tag{2}$$

where K is the consistency index, and n is the flow behaviour index. If n is smaller than one, the fluid is pseudoplastic, therefore the viscosity decreases when the velocity gradient is increased (e.g., fruit juices, sauces). If n exponent is higher than one, the fluid is called dilatant. This type of fluid is less common (e.g., uncooked paste with water and cornstarch, blood) and the viscosity increases when the velocity gradient is increased. If n value is one, the fluid is Newtonian. If the rheological properties of the fluid are determined to develop models in different temperature, according to [9,10], the Equation (2) can be written as expression (3):

$$\mu_{ap} = K(\dot{\gamma})^{n-1}e^{T/T_0} \tag{3}$$

where $\dot{\gamma}$ is the shear rate, T_0 is the reference temperature, and T is the temperature fluid.

If the viscosity value is determined, its determination must be done with an appropriate apparatus, which is related to nature fluid. This measurement devices can be classified into three types: capillary, rotational, and mobile.

Capillary type: this type of viscometer is probably the most used to determine the viscosity, which is measured considering the average circulating flow through a tube as well as the applied pressure. The Hagen-Poiseuille's equation is used to determine the viscosity value [11,12]. These devices are useful to measure great types of fluids. American Society for Testing Materials (ASTM) describe the experimental measure procedure as a function of the considered fluid. The capillary viscometer can be classified in: (i) glass capillary viscometer, which is used to measure Newtonian fluid, (ii) cup viscometer (this type is also called orifice viscometer) and (iii) extrusion viscometer (this is also named piston viscometer). Ostwald viscometer is the basic design of this glass capillary viscometer. Cannon-Fenske, Ubbelohde, FitzSimons, SIL, Atlantic, Zeitfuchs are another similar viscometer [13]. A singular capillary viscometer is the tube of viscometry, which is used to measure the viscosity of suspensions [14]. The cup viscometers are used to measure the fluid properties in the manufacture, process and different applications of dies, paints, and adhesives. Cup viscometers (e.g., Ford, Zahm, Shell, Saybolt, and Furol) are simple and non-expensive rapid methodological devices that are widely used in quality control of Newtonian or near-Newtonian liquids where extreme accuracy is not needed [7]. In this case, the viscosity cannot be determined using the Hagen-Poiseuille equation because the flow analysis is more complex, and the consideration of the emptied time is necessary as parameter to measure the fluency (e.g., Saybolt seconds, Ford seconds). This viscometer type should not be used with non-Newtonian fluid. Finally, the piston viscometer is used to measure the viscosity of fluids very viscous such as molten polymers. This viscometer is made up using a reservoir, which contains the fluid to measure, which is connected to capillary. The fluid crosses the capillary by the force applied due to some sort of piston action.

Rotational type: this type of viscometers is made up by two parts, separated by the fluid to analyse. Depending on the viscometer type, these parts can be: two cylinders (coaxial rotational viscometer), two plates (parallel plates viscometer), and a surface and cone (cone-plate viscometer). In all cases, one of the parts is mobile and the other is fixed. The movement of cylinder, plate or cone generates a shear stress, being determined the viscosity as function of the applied torque to mobile element, rotational speed, radius of the cylinder, plate, or cone; and length of the cylinder or cone. The most famous rotational viscometer is Couette viscometer [15]. The detailed analysis of the equations is shown for coaxial rotational viscometer [16] and for cone-plate and parallel plate viscometer [17].

Mobile type: this type of viscometers operates by the mobility of a mobile part (e.g., sphere, disc, bubble) inside of the fluid that the viscosity is aim to determine. The most famous viscometer of this type is falling-ball viscometer. In this viscometer, the fall time of a sphere is measured to determine the viscosity of the fluid by Stokes equation. The designed viscometer which is described and tested in this research is classified in this type [18].

Looking for Alternatives to Improve the Learning Results by Changing Traditional Methods

When these complex concepts are taught, the inductive learning is more recommended. It is because the student learns the concept, equations or principles through experimental cases [4,5]. The development of experimental tests allows students to assimilate the theory better and reach the learning results at a greater degree. As a final stage, the formulation is deduced for the definition of viscosity. This particular case intends to improve the learning results related to viscosity using an experimental device in which the students firstly check the different behaviour of the fluids as a function of nature, and later, they can assimilate the theoretical expression and concept that command the fluid nature.

The use of the experimentation in different fluids allows students to understand the viscosity concept using easy tools as well as the viscosity variation when the fluid and/or the temperature change.

2.2. Materials and Methods

Previously, to make the proposed viscometer, a group of students proposed the new educational viscometer assisted by the authors of this contribution. The main design constraints of the apparatus were its easy application as well as the obtaining of intuitive results. The obtained values and their analysis will encourage future students to understand the behaviour of the different fluids as a function of their nature. The use of this advice allows students to get an inductive learning, in which the knowledge is provided by experimental activities, reducing the master class hours.

2.2.1. Materials. Apparatus Designed

The viscometer was designed and built by four independent elements (Figure 3): (a) one calibrated cylinder with the inner diameter of 0.092 m and the total height of 1m. The calibrated volume was 6.64×10^{-3} m^3. The primary function of this element was to contain the tested fluid; (b) guided piston, which was made up of two covers. The diameter was near the inner diameter of the element (a). The lower tap had three holes which had an inner diameter of 9×10^{-3} m each one. This cover was connected to the upper tap by three screw bars. The length was 0.5 m. The upper tap whose inner diameter was equal to lower upper guides to lower tap. This is to ensure the parallel and lineal displacement of the lower tap through the cylinder. (c) series of eight calibrated weights to put in the piston and to increase the applied force to the fluid. (d) cylinder with an inner diameter of 0.49 m and a height of 0.50 m. The function of this external cylinder was to maintain the temperature of the inner calibrated cylinder (a). This cylinder will be full of water, which could be at different temperatures using one electrical resistance.

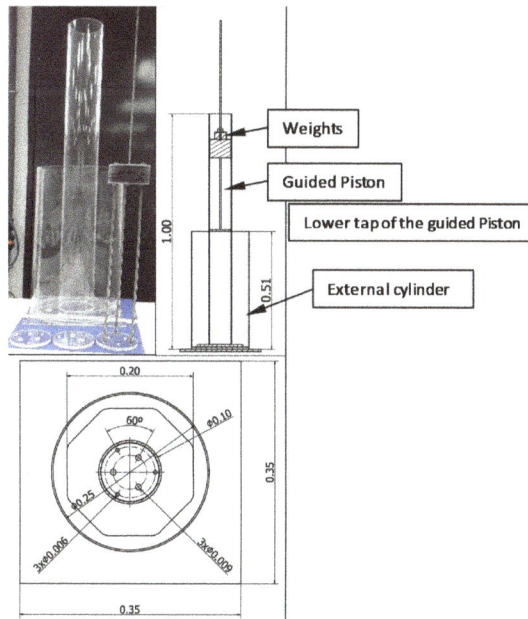

Figure 3. Dimension schemes of the viscometer (m).

The operating of the viscometer was easy. Firstly, the fluid was introduced into the cylinder (a) until the flow level reached a height of 0.43 m. Once the cylinder contained the fluid, the piston was introduced into the cylinder (a). The initial position of the piston was established just above the open surface of the fluid. When the piston was in the initial position considering the selected weight, the piston was left free and was moved by gravity. With the piston in movement, the fluid crossed the lower tap through of the holes, stopping when the piston reached the bottom of the cylinder. In this final position, all fluid was located above of lower tap of the piston. As the measurement had to be repeated, the operator moved the piston to the initial position.

2.2.2. Work Methodology of the Experimental Practice

The objective of this experimentation was the students differentiated the nature of fluids through their behaviour on the device measurement. It was of paramount importance to propose them a methodological strategy to proceed and they were an active part of their learning process about this physical concept of viscosity. Therefore, the proposed methodology for the experimental practice was described in the following points:

(1) Fluid selection: In this first step, the student selected the fluid according to the proposed fluids by the professors. If the fluid was a mixture of different matters, the mixture was made in this step (e.g., the uncooked paste with water and cornstarch). Once the fluid was prepared to be used, the fluid was introduced into in the cylinder, activating the heating to keep the temperature constant during the experiment;

(2) Weights selection: The weights were selected depending on the type of the fluid, since each fluid needs a different value of shear stress (e.g., when the uncooked pate with water and cornstarch is tested needs more weight than water. In the study case, the eight calibrated weights were proposed to be used. The weights (W), which units are kilograms (kg), were 0.130 ($W1$), 0.129 ($W2$), 0.130 ($W3$), 0.259 ($W4$), 0.536 ($W5$), 0.534 ($W6$), 1.074 ($W7$), 1.043 ($W8$), and 0.509 (*Weight WP*). WP corresponded with the piston weight. $W1$, $W2$ and $W3$ are approximately equal to increase the flexibility in the applied force by the piston in the different tests. The selected weight established the shear stress, which depended on the weight and cylindrical surface of the orifices (A). This stress was defined by the Equation (4):

$$\tau = \frac{WT}{A} \tag{4}$$

where WT is the total weight of the piston (own piston weight (WP) plus all considered weights) in N; A is the total cylindrical surface of the orifices in m^2, which was defined by Equation (5):

$$A = N2\pi rh \tag{5}$$

where N is the number of orifices; r is the radius of the orifice in m, and h is the thickness of the tap in m.

(3) Test development: Each analysed fluid was tested under different temperatures (e.g., 20, 30, and 40 °C) and weights (minimum five weights were selected according to tested fluid). For each stage (one fluid and temperature constant), three repetitions were carried out. The time was measured in each repetition and test. Once the data were known, these were managed to determine the average fall velocity (V) and shear stress (τ). When these data were calculated, the experimental results could be drawn, looking at trend line of the fluid (e.g., linear or expotential) and determining its nature;

(4) Guarantee of uniform temperature: The temperature was remained constant to obtain reliable results. This condition was reached through of an immersion bath where the viscometer was introduced. The temperature was remained uniform since two resistances were connected and one temperature sounder measured the temperature value, guaranteeing the uniform value along the test;

(5) Determination of viscosity: the use of the viscometer allowed students to determine the absolute viscosity of the tested fluid using the Newton's equation. If the determination of these values must be calculated, the viscometer had previously been corrected considering the introduction of a geometrical parameter (GP). GP was a coefficient, which was inherent to viscometer and it was related to the manufacturing process. Considering GP, the absolute viscosity was defined using Equation (6):

$$\mu_e = GP\frac{\tau}{V} \tag{6}$$

3. Results

The described results showed an example of the obtained experimental results by the students. These pupils studied the subject Fluids Mechanics, in the second course of Mechanical Engineering Bachelor's Degree. There were one hundred students between 19 and 25 years old. They studied the subject during the academic year 2016/2017.

The tested fluids by the students were water, ketchup, and uncooked paste with water and cornstarch. Their average densities were: 995.71 kg/m^3; 1140.16 kg/m^3; and 1466.19 kg/m^3, respectively, when the temperature was 30 °C.

The development of the described methodology enabled to characterize the different tested fluids. Furthermore, it helped students to understand the fluid behaviour depending on its nature and velocity. The minimum developed tests were fifty-four in each fluid (minimum six test with three repetitions at each tested temperature). The tested temperatures were 19.40 °C, 29.75 °C, and 40.45 °C for the water. When the ketchup was tested, sixty-three tests were done, being the tested temperature 29.50 °C, 40.50 °C, and 46.15 °C. Finally, one hundred seventeen tests were developed for the water-cornstarch mixture for the three temperatures (23.25 °C, 33.25 °C, and 40.00 °C).

Figure 4a–c show the relation between shear stress and velocity in the different developed tests for each fluid, which were obtained by the students. All figures showed the behaviour of each fluid with a good coefficient fit (R^2 is above 0.99 in all cases). Figure 4a shows water's Newtonian condition as well as the viscosity reduction when the temperature increases. This decrease can be observed in the lessening of the slope of the straight line, which crossed near frame's origin.

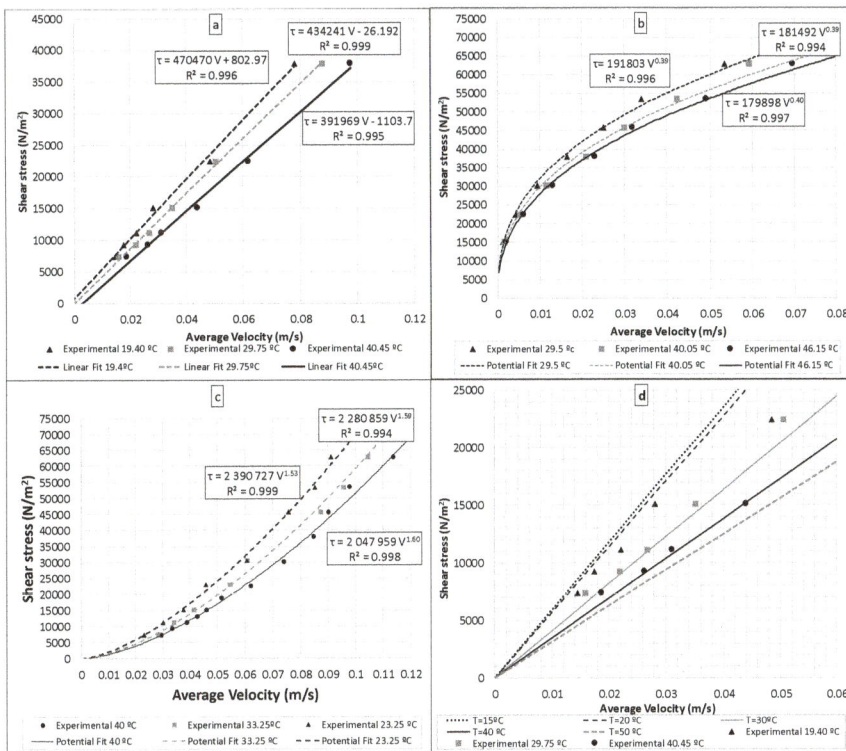

Figure 4. Shear stress vs Average velocity for the water (**a**); ketchup (**b**); corn mixture (**c**); and comparison between theoretical expression and experimental value for water tested (**d**).

Similar results were obtained in the tests with ketchup (Figure 4b). The developed experimental viscometer established the pseudoplastic condition of the ketchup with the obtained result (τ *vs.* V pairs data). As water case, the reduction of the viscosity can be observed when the temperature was increased. The fit showed the exponent value was smaller than one as it was described in Equation (2). The obtained experimental data showed the pseudoplastic condition. The viscosity reduced when the velocity gradient was increased. Figure 4c shows the results for the water-cornstarch mixture. In this case, the viscometer defined the dilatant condition of the fluid, showing the potential fit an exponent greater than one. Also for dilatant fluid, the viscometer was perfectly able to determine the nature fluid, increasing the viscosity value when the gradient velocity increases. This non-Newtonian's behaviour (ketchup and corn starch) was perfectly sensed in the developed tests. This visibility in the behaviour helped the user (students, who carried out the practice) to understand the concept of viscosity.

Figure 4a–c contain the effect of temperature variation. These figures show the proportionality between shear force and velocity with the dynamic viscosity. Figure 4d shows the theoretical expressions for different values of temperature (e.g., 15 °C, 20 °C, 30 °C, 40, and 50 °C), considering the temperature effect on the viscosity.

Finally, the students compared the experimentally obtained values in the water tests with the obtained theoretical expressions for the temperature equal to 19.4 °C, 29.75 °C, and 40.45 °C. This comparison was developed, considering the average value of the regression constant (K) (average value of the slope for each tested temperature in the water) as well a temperature of 20 °C. Figure 4d shows an accurate fit between experimental values and theoretical expressions. This fit enabled to estimate the viscosity of the water for different temperatures. The presented fit can be improved when students increase the number of tests for different temperatures. The increase of test numbers will improve the knowledge of constant value (K). The same way, these equations can also be developed for both non-Newtonians fluids (i.e., ketchup and cornstarch). Finally, the corrector value (*GP*; Equation (6)) of the viscometer was determined for the water. Average *GP* was calculated from experimental results, considering a temperature equal to 40.45 °C. The average *GP* was 1.79×10^{-9} m. Once, this parameter was determined for the temperatures 29.75 and 40.45 °C, the dynamic viscosity can be determined using Equation (6) and experimental data.

The obtained average dynamic viscosity value was 77.9×10^{-5} kg/ms, which had an average error of 5.65% compared with the real dynamic viscosity for the temperature of 30 °C (79.7×10^{-5} kg/ms [6]). When the viscosity value was determined for a temperature of 20°C, the value was 93.6×10^{-5} kg/ms. The average error was 6.53% if the experimental viscosity was compared with real dynamic viscosity value for the temperature of 20 °C (100.5×10^{-5} kg/ms [6]).

This activity proposed interesting learning results, the usefulness of the viscometer to support active methodologies in class of Fluid Mechanics as a complement to the master class were verified both experimentally and by numerous evidences. The simplicity and promptness in the proposed tests enabled the introduction of this experience to encourage the students to be more active and participative during the learning. This activation was promoted by the experimentation of different fluids through the use of the viscometer, using common fluids such as water, ketchup, or cornstarch mixture. The visualization and the fundamental parallel analysis (following the described methodology) to determine the fluid viscosity as well as the representation of the experimental data allowed the students to internalize the concept and the variables on which the viscosity depends (e.g., temperature, shear stress, velocity).

4. Conclusions and Future Applications

The development of this didactical experience caused two positive results. On the one hand, the students worked the transversal competences of teamwork. They practised thinking, designed specific instrumental, and assembled the viscometer with the help of the professors. This design was created to help the teaching staff to re-think the way to teach the first unit of Fluid Mechanics' matter, particularly, the viscosity as well as its properties. The design and assembly of the apparatus joined to the development of the tests checked the appropriate behaviour of the viscometer in the characterization of fluid nature. The results showed the apparatus was able to define the type of fluid: Newtonian, or non-Newtonian (pseudoplastic or dilatant), showing to students the main concepts related to the viscosity by experimental evidence.

A work methodology was proposed to develop the experimental practice, which was designed by the professor in charge of the subject. This methodology enables to reproduce the experimental tests in any subject of Fluid Mechanics and courses. The use of this simple viscometer enabled to determine the fluid nature. The experimental results were fitted as a function of fluid's nature, presenting all regression coefficient values upper than 0.99 (Figure 4a–c). In all cases, the viscometer was sensible to the variation of the temperature. Besides, the experimental dynamic viscosity was estimated for different temperatures of the water. Low error values were obtained when the results are compared with published values in the bibliography.

Future researches will be focused on checking if the use of this apparatus can improve the learning results in the students of Bachelor Degree. Normally, these students have difficulty to understand abstract concepts. These concepts can be understand better if the students develop empirical activities. Currently, the use of this viscometer has been using in the subject 'Fluid Mechanics' that is taught in the second course of Mechanical Engineering Bachelor Degree in the Universitat Politècnica of Valencia. The observed results are encouraging and it is living up the development of the viscosity concept. The future work line is to develop simple apparatus that can be used by the students, improving the learning results and introducing the transversal competences in the students' curricula. Although this research is to show an experimental equipment to develop teaching of the concept viscosity, a first survey were carried out to identify that method the participants would prefer to be taught through. The results showed that the inductive way of teaching was prioritized, since complex concepts were learnt in a natural way. This practice motivated students to continue their learning, synthesizing the contents of the subject, thus reducing the abandonment rate. The innovation improvements will be published in future research related to the improvement of the learning goal.

Author Contributions: Formal analysis, M.P.-S.; Investigation, M.I.P.-S.; Supervision, P.A.L.-J.; Writing—original draft, M.P.-S.; Writing—review & editing, R.G.-S. and P. A.L.-J.

Funding: This research received no external funding.

Conflicts of Interest: The authors declare no conflict of interest.

References

1. Artigue, M. Learning Mathematics in a CAS environment: The genesis of a reflection about instrumentation and the dialectics between technical and conceptual work. *Int. J. Comput. Math. Learn.* **2002**, *7*, 245–274. [CrossRef]

2. Masoliver, G.; Pérez-Sánchez, M.; López-Jiménez, P. Experimental model for non-Newtonian fluid. *Model. Sci. Educ. Learn.* **2017**, *10*, 5–18. [CrossRef]

3. Prince, M. Does active learning work? A review of the research. *J. Eng. Educ.* **2004**, *93*, 223–231. [CrossRef]

4. Andersen, L.; Boud, D.; Cohen, R. Experience-based learning. In *Understanding Adult Education and Training*; Foley, G., Ed.; Allen & Unwin: Crows Nest, Australia, 2000; pp. 225–239.

5. Felder, R.M.; Silverman, L.K. Learning and teaching styles in engineering education. *Eng. Educ.* **1988**, *78*, 674–681.

6. White, F.M. *Fluid Mechanics*, 6th ed.; McGrau-Hill: New York, NY, USA, 2008.

7. Bourne, M. *Food Texture and Viscosity: Concept and Measurement*; Elsevier: Cambridge, MA, USA, 2002.

8. Streeter, V.L. *Mecánica de los Fluidos*; Ediciones del Castillo SA: Madrid, Spain, 1963.

9. Wu, B.; Chen, S. CFD simulation of non Newtonian flow in anaerobic digesters. *Biotechnol. Bioeng.* **2008**, *99*, 700–711. [CrossRef] [PubMed]

10. Bridgeman, J. Computational fluid dynamics modeling of sewage sludge mixing in an anaerobic digester. *J. Adv. Eng. Softw.* **2012**, *44*, 54–62. [CrossRef]

11. Hagen, G. Ueber die Bewegung des Wassers in engen zylindrischen Roehren. *Pogg. Ann.* **1839**, *46*, 423.

12. Poiseuille, J.L.M. Recherches expérimentales sur le mouvement des liquides dans les tubes de très petits diamètres. *Mem. Acad. R. Sci. Inst. Fr. Sci. Math. Phys.* **1846**, *9*, 433–545.

13. Kirk, R.E.; Othmer, D.F. *Encyclopedia of Chemical Technology*; John Wiley & Sons, Inc.: Hoboken, NJ, USA, 2000.

14. Steffe, J.F. *Rheological Methods in Food Process Engineering*; Freeman Press: East Lansing, MI, USA, 1992.

15. Couette, M.M. Etudes sur le frottement des liquides. *Ann. Chim. Phys.* **1890**, *21*, 433–510.

16. Whorlow, R.W. *Rheological Techniques*; Ellis Harwood: Chichester, UK, 1980.

17. Slattery, J.C. Analysis of the cone-plate viscometer. *J. Colloid Sci.* **1961**, *16*, 431–437. [CrossRef]

18. Worsnop, B.L.; Flint, H.T. *Advanced Practical Physics for Students*, 9th ed.; Methuen: London, UK, 1951.

education
sciences

MDPI

Article

"There Is Never a Break": The Hidden Curriculum of Professionalization for Engineering Faculty

Idalis Villanueva [1,*], Taya Carothers [2], Marialuisa Di Stefano [3] and Md. Tarique Hasan Khan [4]

[1] Department of Engineering Education, College of Engineering, Utah State University, Logan, UT 84322, USA
[2] International Office, Advising Services, Northwestern University, Evanston, IL 60208, USA; taya.carothers@northwestern.edu
[3] Department of Teacher Education and Curriculum Studies, University of Massachusetts Amherst, Amherst, MA 01003, USA; marialuisadi@umass.edu
[4] Department of Engineering Education, College of Engineering, Utah State University, Logan, UT 84322, USA; tariquehasan@aggiemail.usu.edu
* Correspondence: idalis.villanueva@usu.edu

Received: 1 August 2018; Accepted: 20 September 2018; Published: 22 September 2018

Abstract: The purpose of this exploratory special issue study was to understand the hidden curriculum (HC), or the unwritten, unofficial, or unintended lessons, around the professionalization of engineering faculty across institutions of higher education. Additionally, how engineering faculty connected the role of HC awareness, emotions, self-efficacy, and self-advocacy concepts was studied. A mixed-method survey was disseminated to 55 engineering faculties across 54 institutions of higher education in the United States. Quantitative questions, which centered around the influences that gender, race, faculty rank, and institutional type played in participants' responses was analyzed using a combination of decision tree analysis with chi-square and correlational analysis. Qualitative questions were analyzed by a combination of tone-, open-, and focused-coding. The findings pointed to the primary roles that gender and institutional type (e.g., Tier 1) played in issues of fulfilling the professional expectations of the field. Furthermore, it was found that HC awareness and emotions and HC awareness and self-efficacy had moderate positive correlations, whereas, compared to self-advocacy, it had weak, negative correlations. Together, the findings point to the complex understandings and intersectional lived realities of many engineering faculty and hopes that through its findings can create awareness of the challenges and obstacles present in these professional environments.

Keywords: hidden curriculum; engineering; faculty; professionalization; mixed-methods

1. Introduction

The goal of this research is to explore how engineering faculty understand hidden curriculum (HC) in higher education settings and their overall reactions and responses to the prevalence of HC in engineering. HC is defined as the unwritten, unofficial, unintended values, lessons, and perspectives that are present in an academic settings and work environments [1–8]. Since the exploration of HC in engineering is very limited [4–6], this special issue exploratory study aims to expand upon the existing knowledge-base about HC to elucidate the mechanisms that faculty use to internalize and communicate their thoughts on this phenomenon.

This study was conducted to help shed light on the lived realities of engineering faculties across several institutional types and is intended for different administrative entities, faculty, and students in Colleges of Engineering, in order to create awareness of the challenges and obstacles that many engineering faculty face. It is the hope of the authors that those who read this manuscript can be inspired to start a conversation about working together to develop healthy and equitable working environments for all.

2. Theoretical Framework

2.1. Hidden Curriculum (HC)

In school settings, curriculum can come in many forms: (a) formal curriculum; (b) informal curriculum; (c) null curriculum; and (d) hidden curriculum [7,8]. The formal curriculum consists of a set of explicitly stated requirements (e.g., rubric) that serve as official guidelines for how to engage with and evaluate the quality of work of students and teachers [2,5–18]. The informal curriculum consists of learning that occurs via personal interactions in the classroom or work spaces [7–9,15–18]. The null curriculum represents the elements that are not taught in a classroom due to mandates from higher authorities, a teacher's lack of knowledge or comfort-level about a topic, or stem from deeply ingrained biases and assumptions about a topic. For example, teaching genetics to introduce topics like evolution continue to be controversial topics that some educators in science classrooms opt to not discuss [19]. Hidden curriculum (HC) represents how particular assumptions, beliefs, values, or attitudes manifests themselves implicitly and inadvertently in schooling, learning, and professional environments [1–18].

HC represents the implicit "attitudes, knowledge, and behaviors, which are conveyed or communicated without aware intent" ([20], p. 125). For example, if an instructor decides to not emphasize a topic for an exam, a student may perceive that this concept is not important to learn. Thus, HC functions at the unconscious, nonverbal spaces of the classroom [7,8] and professional settings [15–18].

Similarly, in the classroom, students do not just learn what is being formally presented in the course but also collect other 'hidden' lessons in the process. It is believed that the human mind can process 80% of explicit information and content in an unconscious manner [7]. Over time, these explicit sources gradually slip beneath the realms of conscious reflection to become a norm that is part of a formalized system [7,8]. Thus, the "space between the official and unofficial, the formal and the informal, the intended and the perceived" ([7], p. 35) becomes the realm where HC lies [1–8].

HC has been used in many disciplines including education, psychology, business, and medicine [9–18] as a strategy to uncover and predict potential issues, that if not attended on time, can lead to dire consequences (e.g., drop-out). While HC traditionally has been tied to negative issues, if used and attended to appropriately, can be used to yield opposite outcomes in students, staff, or faculty (e.g., higher performance; motivated learners and workers) [1–8].

To explore HC, researchers rely on metaphors, vignettes, counter-claims [2,7,11,12,14,16] and ethnographic approaches [14] to robustly analyze the values, perspectives, and beliefs of individuals involved in a learning or working environment. Since HC is contextual and situational, there is also a need to explore these issues from a disciplinary and institutional standpoint as systematic and formalized rules and procedures can vary [1–8]. From a more granular perspective, since HC can be very distinct, it is also important to consider how an individual processes internally these implicit cues [4]. To our understanding, no work has attempted to explore the latter, and in particular for engineering [4].

2.2. HC Mechanisms for Engineering

This work builds upon an initial study from the authors calling for the need to explore more holistically and mechanistically how individuals collect and process these 'hidden' messages in learning and working environments (Villanueva et al., 2018). Earlier work conducted by Authors [4,21] suggested that HC can consist of 16 or more distinct factors to inform the processing of these 'hidden' messages although four of them appear more prevalent: (a) awareness; (b) emotions; (c) self-efficacy; and (d) self-advocacy [4,21]. These will be explained more below.

2.2.1. Awareness of the Presence of HC

Awareness of the presence of HC is an important step for an individual to recognize and reflect upon what is being presented and what is being communicated within their surrounding environment [22–25]. Recent advances in social and cognitive psychology, particularly in the area of metacognition (i.e., an individual's belief about their mental state) [26] has expanded our understanding about what constitutes consciousness [24,25]. The term 'conscious' generally refers to a person directly *seeing*, *knowing* or *feeling* a particular mental content rather than having to indirectly infer upon it [23–25]. Awareness is a sub-component of consciousness where an individual recollects internally an experience and represents it externally (e.g., communication) [26–28]. Depending on the level of representation present, an individual can move into the realm of what is not cognizant (unconscious) or misrepresented (meta-consciousness) [24,25]. Regardless of the level of awareness a person may have about an issue, these cannot be brought up to full consciousness unless they are internalized first.

2.2.2. Emotions to Guide HC Processing

Internalization of an experience typically occurs through an individual's emotions. Emotions assist individuals in narrowing down the variables that are of importance and guide individuals to make decisions for several scenarios [18]. Emotions are important to the learning and socialization process of individuals [29] and serve to explore the influences that several forms of subliminal stimuli can have in their communicated responses [24]. In academic settings, emotions consist of many coordinated processes that involve affective, cognitive, motivational, expressive, and peripheral subsystems that are intertwined [29].

Emotions can be manifested in two forms: (a) valence (positive or negative emotions) or (b) activation (focused or unfocused energy). Positively activated emotions (e.g., enjoyment) may increase reflective processes [29] whereas negatively activated emotions (e.g., anger) may result in low levels of cognitive processing [29].

Finally, emotions contribute to how a person learns, perceives, decides, responds, and problems solve [29]. In the context of HC, emotions can help cue to a person how 'hidden' spaces of expressions, glances, gestures, interest, frustration, and other similar observations become evident within a given context or environment [4,21].

2.2.3. Self-Efficacy Regulates Emotions

At the same time, emotions cannot occur unless a person *believes* that they are able to experience or allow oneself to experience emotions such as joy, anger, pride, etc. [30,31]. Thus, self-efficacy or an individual's beliefs on their ability to ameliorate adverse emotional states [32] is believed to regulate emotions. To our understanding, no work has been conducted to explore how self-efficacy regulates emotions needed to process HC.

2.2.4. Self-Advocacy is Guided by Self-Efficacy

In turn, self-advocacy influences how an individual takes control over their own motivation, behavior, and social environment [33,34]. As such, actions such as self-advocacy or an indication of a person's willingness to take action and speak up about a matter to improve their quality of life [35] cannot occur. To our understanding, no work has been conducted to explore how self-advocacy is guided by the self-efficacies that tie to emotions and HC awareness.

2.2.5. Integrating the Four Factors to Explore HC Mechanisms

Together, the authors posit that these four constructs can serve as a baseline by which to understand more holistically and mechanistically how an individual recognizes, reacts to, and gain abilities to execute control over their own learning or professional environment via HC. To our understanding, no research study has attempted to explore HC mechanisms nor in the context of engineering education. Both elements represent a novel approach towards achieving excellence in engineering education, the focal point of this special issue.

2.3. Professionalization in Engineering: A Type of Formalization of HC

Since HC involves the transmission of implicit messages in learning, teaching, and professional spaces [15–18,36,37], it is integral to understand how aspects of formalization [6,7] could slip beneath the realms of conscious reflection to become a norm [7]. As with any profession, the origins of any career path was once guided by a set of beliefs, values, and attitudes that over time, became a norm [36–38]. These norms can be represented in many ways, but one prevalent form is professionalization [38].

Within academia, the professionalization of faculty in higher education becomes the primary system by which lecturers, professors, and researchers rely on to understand how to navigate their academic environments [38–43]. Abbott [39] suggests that professions, in general, carries three major traditions: (a) traits of professions like cultures (e.g., what professionals in this field represent, and whom they associate with); (b) establishment of a formal education (e.g., licensing and accreditation) [40]; and (c) mechanisms to achieve and maintain a privileged position [41]. Thus, every professional discipline includes a different set of formalized norms and practices that uniquely characterizes their profession [38–41]; engineering is not the exception [44–46].

Professionalization of engineering centers around the intent to constantly adapt to the "ever-changing needs of the market, the emergence of interdisciplinary projects, the increasingly complex social and systemic paradigms" ([44], p. 639). Ironically, the literature suggests that engineering education does not parallel this professional intent as its culture and environments have not changed in decades [45,46] and effects due to these causes have resulted in a stagnant and severe underrepresentation of women and other underrepresented groups [47–50]. Thus, it appears that HC may be present in some of these issues and discrepancies, particularly for teaching and learning spaces where faculty are at the center.

For this work, we will focus primarily on engineering faculty, as these individuals are uniquely positioned in their engineering learning environments to serve as change agents if issues of HC are identified. Additionally, since HC is contextual, it will be important to understand how values, beliefs, and attitudes may change based on the institutional type (e.g., Tier 1, Tier 2) as well as other categorical variables such as race, gender, and faculty ranks (e.g., lecturers, assistant professors, full professors). Thus, this work does not aim to put any processes of professionalization in a negative light but rather through faculties' approvals or disapprovals of the HC presented to them, provide a more rigorous and mechanistic understanding of potential issues that may be present in the education and professionalization of engineering.

3. Methods

The research project presented in this manuscript is part of a more comprehensive and extended mixed-methods research that explored the HC perspectives of engineering undergraduates, graduates, and faculty [4,21]. For this study, we focused solely on engineering faculty, as through their voices, HC can be unveiled and more faculty can be positioned to serve as change agents for their students. All items in this study were conducted by adhering to ethical standards and treatment of human subjects as required by the Institutional Review Board (IRB) of the home institution of the first author.

3.1. Research Design

This work uses a complex, mixed-method experimental intervention design [51,52] that incorporates a quasi-experimental quantitative design with qualitative data that is combined in a convergent manner. For this study, integration of the qualitative and quantitative findings of the design occurred in the data analysis and interpretation phases of the work [53]. Additionally, a primary emphasis was placed on the quantitative elements of the study (denoted by the higher caps, 'QUAN') and a secondary emphasis was given to the qualitative components of the work (denoted by the lower caps, 'qual') [51,52].

3.2. Research Questions

The three central hypotheses (H1, H2, and H3) for the QUAN elements of the study are:

H1. *Engineering faculty will recognize HC in their fields regardless of race, gender, rank, and institutional type.*

Rationale: Given that the literature points to the relatively stagnant culture of engineering [44–50], it is possible that the cultural norms, beliefs, values, and attitudes continue to pervade regardless of other intersectional identities (e.g., race, gender).

H2. *Changes among institutional types (e.g., Tier 1 compared to Tier 2) will result in a decreased recognition of HC mechanisms among engineering faculty, regardless of race and gender, when counter-claims of the field of engineering is presented.*

Rationale: Within institutions where tenure is a primary focus [38–41], professional elements such as research and teaching may make faculty more prone to experience the influences of HC.

H3. *Changes among professional faculty ranks (e.g., lecturer versus associate professor), regardless of race, gender, and institutional type will result in an increased presence of HC mechanisms among engineering faculty when counter-claims of the field of engineering is presented.*

Rationale: In parallel to the institutions' research focus, faculty responsibilities may be more geared towards research [38–41] and these focuses on their professions may lead to a higher internalization of HC.

The main research question (RQ) used for the qual portions of this work was:

RQ1. *What are the central messages around the professionalization of engineering that the faculty convey?*

3.3. Participants

The participants were 55 engineering faculty, out of 393 total participants that included undergraduates, graduates, postdocs, and faculty [4,21], working or studying across 54 higher education institutions in the United States and Puerto Rico. The breakdown of the institutional types assessed and regions are summarized in Table 1. Note that while Carnegie classification was used as a guide to label the institutional types [54] some were clustered (e.g., community colleges) to allow for statistical comparisons among the groups. Additionally, to protect the anonymity of the participants, no institution name was included.

Table 1. Summary of institutional types where engineering faculty were employed.

Institutional Type (Tier Label)	Carnegie Classification Description	% Faculty Participants
R1 (Tier 1)	Doctoral universities: highest research activity	48%
R2 (Tier 2)	Doctoral universities: higher research activity	13%
R3 (Tier 3)	Doctoral universities: moderate research activity	9%
M1-M3 (Tier 4)	Master's Colleges and University: All Programs Sizes	15%
B1-B3 (Tier 5)	Bachelor Colleges and Community Colleges: All Program Sizes and Types	15%

Participants were recruited using purposeful sampling [55,56] through email and posts on social media channels connected to engineering professional organizations (e.g., Society for Women Engineers, Society of Hispanic Professional Engineers, and American Society of Mechanical Engineers) as well as LinkedIn and Facebook. The inclusion criteria were: (a) current engineering faculty member at a higher institution in the United States; and (b) response to our call. The exclusion criteria were: (a) retired faculty in engineering; (b) engineering faculty who were not in an institution of higher education in the United States; and (c) practicing engineers in industry. For the latter, we had some international individuals and engineers from industry who responded to the survey but, due to the nature and scope of the work, these were excluded from analysis.

A summary of the ranks (e.g., lecturers, assistant professors) of the engineering faculty participants are described in Table 2. Note that adjuncts and lecturers were considered one type of rank as their primary roles are teaching and not research. Furthermore, a racial and gender breakdown of the engineering faculty participants are summarized in Tables 3 and 4, respectively, as we were interested in understanding if individual intersectional identities based on race and gender could influence these HC mechanisms differently. Note that some of the faculty self-reported as being multi-racial as shown in Table 3.

Table 2. Summary of ranks of engineering faculty.

Rank		
Adjunct and Lecturers	16	29%
Assistant Professors	15	27%
Associate Professors	11	20%
Full Professor	13	24%
Total	**55**	

Table 3. Summary of race of engineering faculty.

Race		
White	39	60%
Hispanic	10	15%
Asian	8	12%
Black/African American	3	5%
Two or more	5	8%
Total	**65**	

Table 4. Summary of gender of engineering faculty.

Gender		
Female	28	51%
Male	26	47%
Non-binary/third gender	1	2%
Total	**55**	

3.4. Survey Items

The survey items to be discussed in this manuscript are part of a larger survey, whose validation process was discussed elsewhere [4,21]. The validated instrument consists of 10 demographic questions, 24 Likert-scale items (six items per sub-scale: awareness, emotions, self-efficacy, and self-advocacy) and five qualitative questions about participants' views about engineering and HC in engineering and a video vignette. The full instrument has a Cronbach alpha score of 0.70 and sub-scale scores of 0.70 for HC awareness, 0.71 for emotions, 0.82 for self-efficacy, and 0.84 for self-advocacy [21].

The video vignette was created from a systematic review of the engineering education literature before its creation and inclusion in the survey as described elsewhere [4,21]. To summarize the video briefly, the content exposed participants to the dynamics involved during an engineering course preparation by two instructors and the interactions in engineering classrooms between students and these two faculty members. Video elicitation was used as a social science technique to garner participant awareness about a topic [21,57]. In this video, issues of gender dynamics and race were highlighted as the literature suggests a severe underrepresentation of these groups in engineering [44–50]. Furthermore, it is important to mention that while the video may have appeared biased at a glance, the video was representative of a systematic synthesis of the literature (written in script form) of the primary cultures and environments of engineering education [44–50].

Since rigorous analysis of HC requires strategies built upon counter-claims [7], the topics and scopes selected for the video were deemed appropriate to the scope of the work and its fictitious nature was indicated to the participants. The video also went through a rigorous process of validation before it was added to the instrument [21] and its final Cronbach alpha score was found to be 0.87 across the original 393 participants [21]. In a similar vein, the HC statements selected for this instrument represented the video closely as well as the systematic review of the engineering education literature [4,21,44–50]. The full instrument and video links are expected to be released in the near future.

Based upon prior findings and throughout the survey validation process, the authors found that many participants could not recognize HC [21]. As such, it was deemed necessary to carefully place the items of the survey in a way that participants could recognize and reflect upon the HC sub-scales to minimize any potential "mental shortcuts" they may have used to make sense of a new concept or phenomenon ([58], p. 4). Additionally, since framing has an influence over the interpretation of meanings and its connections to ideas and beliefs [58], the research team wanted to make sure its placement would minimize potential variations in the understanding of this phenomenon amongst a diverse set of populations [21].

To summarize and for the purpose of this special issue exploratory study, the survey participants were part of the original participants needed for this survey validation. Participants who took this version of the survey saw the following sequence: (a) demographic questions; (b) qualitative questions; (c) video vignette; (d) video character question; (e) HC awareness questions; (f) emotions questions; (g) self-efficacy questions; and (h) self-advocacy questions. This study focuses primarily on participants' responses to the last four sub-scales and the qualitative questions.

Since HC requires that participants recognize the phenomenon [4,21], a definition of HC was provided based upon what has been identified in the literature [1–8]. Furthermore, to facilitate faculty participants' understanding of HC, the terms "examples", "agree", "disagree", and "statement" were used in the framing of the question to convey to participants that they had a choice and opinion that was valued in their responses. A description of one of the sub-scales (HC awareness) is included in Table 5.

Table 5. HC assumption statements used for this study.

Instructions:
Hidden curriculum (HC) refers to the unwritten, unofficial, and often unintended lessons, values, and perspectives learned in an academic environment. We identified the following six examples of the hidden curriculum in this video. Read each statement. Do you agree or disagree with each statement (**Yes = I agree, this is HC; No = I disagree, this is not HC**)?

HC #1	Senior faculty in engineering (e.g., tenured professor) deserve higher status, voice, and have more influence than engineering junior faculty.
HC #2	The ultimate goal of an engineering degree is to get a well-paying job.
HC #3	Engineering education is harder, more time-consuming, and expensive because it has a direct impact on safety.
HC #4	Not everyone can be an engineer.
HC #5	To belong to the engineering community, your personality must fit in with everyone else's (e.g., technically-driven, efficient, and assertive)
HC #6	Engineering instructors care more about the technical concepts and equations rather than the individual student's success.

Note that for the other sub-scales (emotions, self-efficacy, and self-advocacy), the same six HC statements were used in an attempt to identify potential associations between the four factors. A summary of the presentation of the other sub-scales is provided in Table 6.

Table 6. Modified summary of instructions and Likert-scale descriptions for the other sub-scales in the instrument for this special issue study.

Sub-Scales	Instructions	Likert Scale Response Description
Emotions	What **emotion** would best describe your overall reaction to each statement (**choose from the list**)? Is this overall emotion **positive** or **negative** for you? You can also indicate if this emotion is both positive and negative, or neither one nor the other.	'1'—I felt positive about this statement '2'—I felt negative about this statement '3'—I felt both positive and negative about this statement '4'—I felt neither positive, no negative about this statement
Self-efficacy	**Self-efficacy** is your belief in your ability to succeed in specific situations or accomplish a task. Rate your **confidence** (self-efficacy) in succeeding if placed in a similar situation to the statements provided.	'1'—I am not at all confident that I can succeed in a situation similar to this '3'—I am somewhat confident that I can succeed in a situation similar to this '5'—I am very confident that I can succeed in a situation similar to this
Self-Advocacy	**Self-advocacy** is the ability to speak or act on your own behalf to improve your quality of life, effect personal change, or correct inequalities. Rate your **willingness** to **advocate** for yourself if placed in a similar situation to the statements provided.	'1'—I am not at all willing to advocate for myself in situation similar to this '3'—I am somewhat willing to advocate for myself in situation similar to this '5'—I am very willing to advocate for myself in situation similar to this

For the emotions sub-scale, participants were provided with a list of 17 emotions as recommended by Pekrun and colleagues [29] and DeCuir-Gunby and colleagues [59]. Additionally, one of the valence dimensions (positive/negative) was asked. The results of this study focus on the latter (positive/negative valence). Also, participants were given the choice to choose "neither" or "both" to convey the possible and complex nature of emotions to different contexts and situations [29]. For the self-efficacy sub-scale, participants were asked to rate their confidence in their ability to succeed if placed in a similar situation to the provided HC statement. Terms like "confidence" and "success" were emphasized to convey the definition provided as well as help them contextualize that HC may guide coping strategies based upon their individual beliefs and perceptions. Finally, for the self-advocacy sub-scale, terms like "willingness" and "advocate" were used to convey to participants that they are in full control of their actions and approaches to situations similar to the provided example HC statements.

3.5. Data Preparation and Analysis

3.5.1. QUAN Data Preparation

Pre-processing of data was first required to make it analyzable and to account for any missing values or text entries. All the blank spaces were filled by a category labeled as "No Response (NR)" accompanied by a numerical value. In a similar fashion, all other categorical responses, including demographic information (e.g., gender, race) were assigned a numerical value as shown in Table 7.

Table 7. Representative categorical participant responses that were assigned numerical values.

Gender	Race	Institution Type	Faculty Rank	HC Awareness	Emotions	Self-Efficacy and Self-Advocacy
Female = 1	White =1	Tier 1 =1	Adjuncts/	No = 0	Negative = 0	'1' = 1
Male = 2	Hispanic = 2	Tier 2 = 2	Lecturers = 1	Yes = 1	Positive = 1	'2' = 2
Third Gender = 3	Black = 3	Tier 3 = 3	Assistant Professor = 2	NR = 2 (No response)	Neither = 2	'3' = 3
	Asian = 4	Tier 4 = 4	Associate Professor = 3		Both = 3	'4' = 4
		Tier 5 = 5	Full Professor = 4		NR = 4 (No response)	'5' = 5

3.5.2. Quantitative Data Analysis: QUAN

Advanced Machine Learning: Decision Tree Analysis

For quantitative analysis, the research team opted to use a new approach in their lab: advanced machine learning (AML). AML allows researchers to take a wide array of responses and using algorithms, develop an automated and unbiased way to identify the likelihood of an influence on participants' response [60]. For similar study designs to ours, AML has used small and extendable (not a static or fixed) datasets to help classify what factors and sub-factors are predominant among participant responses [60].

It is important to note that although the dataset consists of only 55 participants, each participant yielded responses to four factors (awareness, emotions, self-efficacy, and self-advocacy) that were classified as six sub-factors (HC statements) and whose variations in sub-factor response (e.g., valence options; lists; Likert-scale entries) were different. As such, the nature of our dataset can be limited in its analysis through traditional statistical methods as oftentimes, these approaches are deficient in their ability to uncover the inter-relationships of varied dependent and independent variables [61]. Furthermore, through this technique, the researchers could consider in an unbiased and automated manner, how a wide array of intersectional demographics contributed to participants' responses. Hence, at a minimum, the dataset consisted of 24 independent factors/variables/features to analyze (e.g., 24 × 55 participants = 1320 minimum data points).

Within AML, a decision tree (DT) analysis was used [62–64]. DT was used in this study to understand the relational aspects between the factors, sub-factors, and specified categorical variables (i.e., gender, race, faculty rank, institution type). Decision trees include a feature (attribute), a link to each feature (branch), and a representative decision (rule) that represents an outcome (categorical or continuous value) [62–64]. For this study, a classification and regression tree (CART) algorithm was used to divide the factors, sub-factors, and variables to identify decision rules that can denote the relative importance or predominant influence of a variable. For CART, HC awareness consisted of six labels (v1–v6). Emotions had six labels starting from v7 to v12. In a similar fashion, variables are arranged for self-efficacy (v13–v18) and self-advocacy (v19–v24). A representative decision tree is shown in Figure 1 where the influence of gender over the factor HC awareness was found for the third (v3) and fourth (v4) HC example statements from Table 5. This same process was repeated for all the factors.

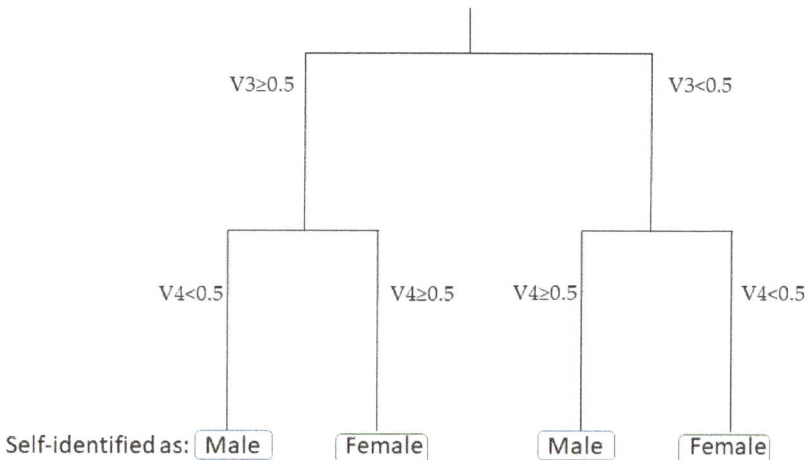

Figure 1. Decision tree to understand the influence of gender over HC awareness.

The DT example provided in Figure 1 shows how a decision tree works. In this example, a decision tree was used to identify the influence of gender on participants' responses to HC awareness. This yielded four leaves, known as decision rules, which for this example would be:

Rule 1: When (V3 ≥ 0.5 = Yes) AND (V4 ≥ 0.5 = Yes) THEN the Gender is Male
Rule 2: When (V3 ≥ 0.5 = Yes) AND (V4 < 0.5 = No) THEN the Gender is Female

Rule 3: When (V3 < 0.5 = No) AND (V4 $\geq$ 0.5 = Yes) THEN the Gender is Female

Rule 4: When (V3 < 0.5 = No) AND (V4 < 0.5 = No) THEN the Gender is Male

These rules are then given a percentage value, which represents the predominance of the influence that a factor, sub-factor, or categorical variable had in participants' responses. In this example, 2.7% of the influence for the third HC statement (*"Engineering education is harder, more time-consuming, and expensive because it has a direct impact on safety"*) presented predominant influences from participants who self-identified as male. The percentage breakdown of these rules are included in the results section of the manuscript.

Statistical Analysis

Informed upon the AML findings on the predominance of an influence of a given factor, sub-factor, or categorical variable, frequencies of responses were tabulated among the participants. These frequency counts were used to conduct a test of independence via chi-square analysis for the categories of race, gender, faculty rank, and institutional type.

Additionally, among the factors, association/correlation analysis was conducted on the R-squared values for each categorical response to find potential relationships. A best-fit line was run on each factor (e.g., HC awareness, emotions, self-efficacy, and self-advocacy) to see if there were positive or negative trends between each factor. Correlation coefficient values of 0.3 or less indicate weak correlations; values close to 0.5 indicated moderate relationships and above 0.7 indicate strong relationships. Positive or negative signs portrayed the directionality of the correlation [61].

3.5.3. Qualitative Data Analysis: Qual

For the qualitative analysis of the survey, we used a method advanced by Loftland and colleagues [55] for social science framing. First, the researchers read through the written responses of all participants to understand the general tone of the responses. Next, each response was assessed line-by-line through open-coding [56] to identify themes and any possible question misunderstandings. Then, each written response was read a third time and focused-coding was conducted to identify distinct categories. These were then analyzed once more for interpretation, categorized into themes, connected theoretically, and the fit for each theme was assessed. For our analysis, added attention was paid to what could be identified or interpreted as an example of HC in from participants' personal narratives or additional experiences.

3.6. Researchers' Positionality

Each researcher in this work comes from a different background and has developed an epistemology specific to the fields of engineering (faculty and postdoctoral fellow), curriculum and instruction (faculty), and social science (assistant administrator). These perspectives helps provide a holistic understanding of the HC explored for this population. Each researcher has been privy to the influences that HC could have in their professional formation, as well as has experienced both the negative and positive influences of such. It is the hope of the research team that, by uncovering this HC, power dynamics can become more democratized within this field [4].

4. Results

4.1. Quantitative Findings: QUAN

4.1.1. Presence of HC Mechanisms among Engineering Faculty

To answer the first hypothesis (H1), a frequency count of faculty responses were measured for the four sub-factors (awareness, emotions, self-efficacy, and self-advocacy). For HC awareness, all faculty were approximately equal among what they considered was an HC in their field or not (Table 8). Among the statements identified as HC in engineering, were HC #1 (*"Senior faculty in engineering (e.g., tenured professor) deserve higher status, voice, and have more influence than engineering junior faculty"*), HC #4 (*"Not everyone can be an engineer"*) and HC #6 (*"Engineering instructors care more about the technical concepts and equations rather than the individual student's success"*). Among the statements that faculty regarded as not being hidden curriculum were HC #2 (*"The ultimate goal of an engineering degree is to get a well-paying job"*), HC #3 (*"Engineering education is harder, more time-consuming, and expensive because it has a direct impact on safety"*), and HC #5 (*"To belong to the engineering community, your personality must fit in with everyone else's (e.g., technically-driven, efficient, and assertive)"*).

Table 8. Frequency counts of faculty participants' agreement or disagreement that the statements were an HC in engineering.

Participant Response	HC #1	HC #2	HC #3	HC #4	HC #5	HC #6	SUMS
Yes. I agree this is HC	31	17	27	38	17	31	161
No. I disagree this is HC	22	36	26	16	37	22	159

A frequency count of the emotional valence (e.g., positive/negative) was tabulated for the participants as summarized in Table 9. While most faculty participants reported having to experience negative emotional valence to the HC statements, a combination of positive emotions or neither was also self-reported. Among the statements, HC #6 (*"Engineering instructors care more about the technical concepts and equations rather than the individual student's success"*) yielded the most negative emotions while HC #3 (*"Engineering education is harder, more time-consuming, and expensive because it has a direct impact on safety"*) resulted in the most positive emotions.

Table 9. Frequency counts of faculty participants' self-reported emotional valence response to the HC statements about engineering.

Emotional Valence	HC #1	HC #2	HC #3	HC #4	HC #5	HC #6	SUMS
Positive	5	12	18	10	7	7	59
Negative	31	25	15	20	32	33	156
Both	11	11	10	10	9	10	61
Neither	8	6	8	13	5	3	43

A frequency count of self-efficacy (e.g., confidence in the ability to succeed despite the HC) was tabulated for the participants as summarized in Table 10. Most engineering faculty participants reported feeling very confident about their success despite the possible presence of similar scenarios to the HC statements. Particularly, HC #4 (*"Not everyone can be an engineer"*) yielded the highest level of self-efficacy among participants. Among the lowest reported levels of self-efficacy, HC #1 (*"Senior faculty in engineering (e.g., tenured professor) deserve higher status, voice, and have more influence than engineering junior faculty"*) and HC #2 (*"The ultimate goal of an engineering degree is to get a well-paying job"*) were considered as having elements that would not enable them to succeed in these contexts.

Table 10. Frequency counts of faculty participants' self-reported self-efficacy to the HC statements about engineering.

Self-Efficacy	HC #1	HC #2	HC #3	HC #4	HC #5	HC #6	SUMS
'1' (not at all confident)	7	7	5	4	6	6	35
'2' (mainly not confident)	10	5	7	10	8	7	47
'3' (somewhat confident)	16	15	12	12	13	13	81
'4' (moderately confident)	13	16	16	11	12	12	80
'5' (very confident)	9	12	15	18	16	17	87

A frequency count of self-advocacy (e.g., willingness to advocate for HC) was tabulated for the participants as summarized in Table 11. For the most part, engineering faculty were somewhat willing to advocate for HC, particularly for HC #1 (*"Senior faculty in engineering (e.g., tenured professor) deserve higher status, voice, and have more influence than engineering junior faculty"*) and HC #4 (*Not everyone can be an engineer"*). Among the statements where the faculty were not at all willing to advocate was HC #3 (*"Engineering education is harder, more time-consuming, and expensive because it has a direct impact on safety"*) and HC #2 (*"The ultimate goal of an engineering degree is to get a well-paying job"*).

Table 11. Frequency counts of faculty participants' self-reported self-efficacy to the HC statements about engineering.

Self-Advocacy	HC #1	HC #2	HC #3	HC #4	HC #5	HC #6	SUMS
'1' (not at all willing)	9	3	2	4	4	7	29
'2' (mainly not willing)	5	8	6	4	9	7	39
'3' (somewhat willing)	15	13	14	15	11	14	82
'4' (moderately willing)	12	15	15	13	13	14	82
'5' (very willing)	11	13	13	15	13	9	74

4.1.2. Predominance of HC Factors, Sub-Factors, and Categorical Variables Based on AML and DT

To answer the second and third hypothesis (H2 and H3), the predominance of an influence among participant responses were calculated using AML and DT, through relative importance of variable algorithms [62–64]. A summary of the predominance of an influence (in the form of percentages) is presented in Table 12.

Table 12. Summary of the predominant influence (in the form of percentage) of the factors and sub-factors identified from the engineering faculty participant responses due to categorical variable.

Factor	Sub-Factor	Race	Gender	Faculty Rank	Institutional Type
Awareness	HC #1	33.5%	10.6%	26.8%	8.5%
	HC #2	26.5%	10.3%	3.1%	26.7%
	HC #3	9.5%	2.7%	53.7%	6.8%
	HC #4	7.1%	9.5%	1.6%	0.2%
	HC #5	1.1%	54.0%	1.0%	8.4%
	HC #6	22.3%	12.9%	13.8%	49.3%
Emotions	HC #1	75.0%	9.6%	13.0%	1.0%
	HC #2	7.9%	0.2%	19.3%	35.9%
	HC #3	3.0%	1.6%	9.8%	4.0%
	HC #4	6.2%	22.9%	1.6%	47.1%
	HC #5	1.7%	11.5%	2.3%	6.6%
	HC #6	6.2%	54.1%	54.1%	5.4%
Self-Efficacy	HC #1	11.4%	2.2%	1.4%	14.1%
	HC #2	9.1%	7.1%	44.3%	8.7%
	HC #3	42.3%	1.1%	37.1%	1.2%
	HC #4	33.6%	6.7%	1.5%	61.1%
	HC #5	1.4%	5.7%	5.5%	1.8%
	HC #6	2.2%	77.4%	10.2%	13.1%
Self-Advocacy	HC #1	12.8%	21.0%	11.7%	34.0%
	HC #2	11.2%	3.1%	4.9%	11.0%
	HC #3	12.5%	13.4%	13.9%	4.2%
	HC #4	49.8%	20.9%	17.3%	22.1%
	HC #5	12.0%	7.1%	49.2%	4.7%
	HC #6	1.7%	34.6%	3.1%	24.0%

The findings suggested that among the categorical variables, gender showed the highest influence among participant responses regardless of factor and sub-factor. For gender, the primary influences were found on HC #5 (*"To belong to the engineering community, your personality must fit in with everyone else's (e.g., technically-driven, efficient, and assertive)"*) and HC #6 (*"Engineering instructors care more about the technical concepts and equations rather than the individual student's success*"). We also found that for all engineering faculty, closely similar trends of predominance was found for HC awareness and emotion across the six statements. These latter trends did not parallel faculty responses to the self-efficacy and self-advocacy items.

Among the highest influences for race found was HC #5 ("To belong to the engineering community, your personality must fit in with everyone else's (e.g., technically-driven, efficient, and assertive)") followed by HC #3 ("Engineering education is harder, more time-consuming, and expensive because it has a direct impact on safety"). For faculty rank, the primary predominant influences were found for HC #6 followed by HC #3. For institutional type, main influences were found for HC #4 ("Not everyone can be an engineer") and HC #6.

From the AML findings, a chi-square analysis was conducted between the pairings of the four categorical variables (gender, race, institutional type, and faculty rank). Findings revealed that gender and institutional types showed the greatest relationships overall ($\chi^2 = 18.422$; df = 8; $p < 0.05$). To further examine these two relationships and since Tier 1 had a higher percentage of respondents, these values were removed to identify the relationships of gender to the other institutional types. The same process was repeated for the other institutional types until a ranking in influence (denoted by p-values under 0.05) was measured. Findings confirmed the predominant influence of gender was found among the Tier 1 institutions (doctoral-granting institutions with large research programs) followed by close second (Tier 4; masters-granting institutions; $p < 0.05$). Trailing behind were Tier 2, Tier 3, and Tier 5 institutions, respectively. For all other relationships (e.g., faculty rank versus institutional type; race versus institutional type; gender versus race; race versus faculty rank, etc.), no statistically significant relationships were found.

Among the factors, regression analysis was conducted between HC awareness and self-efficacy and HC awareness and emotions. Results were found to have a weak or moderate positive correlation ($r = 0.28$ for self-efficacy; $r = 0.42$ for emotions). HC awareness and self-advocacy revealed a weak negative correlation ($r = -0.31$).

4.2. Qualitative Findings: Qual

To answer the research question, a qualitative analysis of participants' written responses was conducted. These resulted in three recurrent themes: (a) professional expectations for engineering faculty; (b) sources of professional expectations; and (c) consequences of meeting professional expectations in engineering.

4.2.1. Professional Expectations for Engineering Faculty

Findings suggested that teaching, service, and research were the top categories that respondents listed for what expectations they felt were placed on them as faculty. In addition, all faculty described expectations of time commitment and availability as being important. Among the responses, faculty described the expectation of being "able to work whenever needed (nights, weekends, holidays)" (Respondent 41, Full Professor, Tier 5, Male, Hispanic), or "to have all time outside of class to be free and uncommitted" (Respondent 12, Assistant Professor, Tier 5, Male, Hispanic) to be at odds with their work/life balance (i.e., "high productivity, no work-life balance" (Respondent 28, Lecturer, Tier 5, Male, White)).

Broader educational goals that faculty were expected to deliver included providing a high-quality engineering education, conferring credentials only to the deserving students, and meeting the Accreditation Board of Engineering and Technology (ABET) criteria [65]. More specific qualities that engineering faculty were expected to have included problem-solving, attention to detail, good communication skills, ability to work in a team, and being self-critical.

Being able to solve an equation is good, but what is more important is to be able to see if
the answer is correct or not, like finding a negative resistance in a circuit problem can be
mathematically correct but practically inconsistent with real circuit. Self-criticism should be
taught more. This is what I try in my classes.

(Respondent 18, Lecturer, Tier 1, Male, White)

Two faculty commented that a professional expectation is to "not ask many questions"
(Respondent 19, Full Professor, Tier 1, Female, White) and "not create any problems" (Respondent 5,
Associate Professor, Tier 5, Male, Hispanic) and that "mantras such as 'engineers provide solutions,
not problems' " (Respondent 33, Full Professor, Tier 5, Female, White) are valued more among
engineering departments.

Other faculty elaborated that among professional expectations, there are elements of advocacy and
allies that are needed, especially when considering the experiences of women and underrepresented
groups in engineering:

I am a woman and an immigrant. I am a first generation student. I often have not seen
students and colleagues respect me. After I stood up for myself things have started to change.

(Respondent 46, Associate Professor, Tier 1, Female, White)

Other important features are having the right networks/allies to get stuff done, as well as
identifying allies of the majority group to speak up as well. When a woman stands up and
talks about gender bias issues being a problem, it's often not listened to as much as a man
doing it.

(Respondent 48, Full Professor, Tier 1, Male, White)

4.2.2. Sources of Professional Expectations

Survey respondents identified several origins to these professional expectations but most notably
from students, peers, and themselves. Specifically, when describing faculties' impression of student
expectations, some respondents elaborated on their role in helping uncover HC in the classroom:

"I may be the only influence on this topic [hidden curriculum] the students are exposed
to regularly"

(Respondent 34, Assistant Professor, Tier 5, Male, White)

Personally, I don't think many of these . . . [HC statements] are truly hidden curriculum,
at least in my classes I try to explain lessons whether they are part of mainstream engineering
education or not.

(Respondent 28, Lecturer, Tier 5, Male, White)

Others, on the other hand, indicated that their intersectional identity and role could result in
an opposite response from students in the classroom:

" . . . as a faculty [. . .] students are less respectful of me due to my race/nationality/color
of skin/religion..."

(Respondent 36, Assistant Professor, Tier 1, Female, Black)

"In an effort to be professional and polite there are times that I let it [students disrespecting
me] slide and regret it later . . . "

(Respondent 23, Associate Professor, Tier 1, Female, White)

Another theme that emerged was around how professional expectations from peers related to the
intersectional identities of the faculty. One respondent said:

"As a woman in engineering, service is also expected of me more than my male colleagues. My female colleagues and I get asked constantly to do service, while our male colleagues are rarely asked (or, when they are on a committee, have no reservation about saying they are 'too busy' and 'having the women to do work')"

(Respondent 47, Assistant Professor, Tier 1, Female, Black)

Others faculty indicated that they too recognize their roles as being a source of encouragement and mentorship for underrepresented peers:

As I mentioned before, we are still doing things the way they were done 100 years ago. Maybe this is why our enrollment of women and minority groups continues to be so low. I made an effort to recruit minority and women faculty. I mentored them and was successful in helping them to earn tenure. In turn, they have attracted a more diverse group of students. So, I have seen how paying attention to diversity pays off!

(Respondent 32, Associate Professor, Tier 5, Male, White)

Keep on doing what you're doing [referring to a main character in the video]! Although maybe do decrease your service level as an Asst. Prof. And go ahead and include questions on diverse engineers on the exam. It's important.

(Respondent 49, Full Professor, Tier 1, Female, White)

4.2.3. Consequences of Meeting Professional Expectations in Engineering

Several respondents expressed that some of the consequences they experienced while meeting professional expectations in engineering lied in trying to keep a work/life balance: "I could work 24/7 and still never be caught up" (Respondent 47, Assistant Professor, Tier 1, Female, Black); "there is never a break" (Respondent 44, Full Professor, Tier 1, Female, White); and "all areas of concern are unrealistic even for the most dedicated faculty" (Respondent 17, Assistant Professor, Tier 1, Female, White).

Thirty respondents stated that meeting the expectations of being a faculty member led to personal consequences, such as exhaustion and frustration. On the other hand, four faculty members elaborated that they do not feel meeting expectations should be exhausting because they love their jobs, suggesting that if faculty love their job they should not feel exhausted about meeting expectations. Among the participants that referenced reasons why meeting job expectations in engineering can be exhausting included confusion and pressures regarding grading, fulfilling administrative duties, responding to service requests and navigating an outdated system in engineering education. One respondent stated, "trying to be academically rigorous and supportive to the students and somehow getting research done at the same time is completely exhausting" (Respondent 23, Associate Professor, Tier 1, Female, White). Some respondents suggested that the exhaustion derives from their interactions with students who are underprepared and underperforming. One faculty wrote that "many students do not work hard enough" (Respondent 29, Assistant Professor, Tier 2, Female, White) and several faculty members elaborated on the academic rigor of the engineering curriculum. Three respondents referred to the competitive research environment coupled with the pressure to conduct innovative research.

Regarding the "frustrations" experienced by many engineering faculty participants, the challenges with navigating an outdated system in engineering education and the powerlessness that many feel for creating sustainable changes in engineering exist:

I am so completely frustrated with the older generation of professors. They're mostly white, almost all male, and completely uninterested in how to teach better. They refuse to see student struggles as anything but laziness. I was first generation college student - I know what that's like. I get good evaluations, I advise and mentor students, and I bend over backward to make sure their experience is good. But because it feels like I get no credit for it, and previous attempts to advocate for change go nowhere, I don't feel like I can ever make a difference outside of the students I directly interact with.

(Respondent 25, Tier 1, Associate Professor, Female, White)

I can advocate for this in my classrooms, but not among the all-male, mostly white faculty.

(Respondent 47, Assistant Professor, Tier 1, Female, Black)

For others, these frustrations became their personal sources of motivation for changing the status quo in engineering:

It is difficult to do the right thing. The first step is rejecting the status quo. My anger and shame about the "good ole boys club" in engineering has helped me to fight against it!

(Respondent 32, Associate Professor, Tier 5, Male, White)

5. Discussion

5.1. QUAN Findings

In this study, it was found that engineering faculty were able to recognize the presence of hidden curriculum for some scenarios and not for others. For those statements that predominantly identified as HC, faculty recognized that the high demands for engineering (HC #3) and the professional demands of their field, may influence how engineering instructors are perceived regarding their care to students' success (HC #6). This parallels to what the engineering education literature states are the difficult nature of the field [44–50]. Additionally, regarding instructor care, it appears that there may be a level of awareness from the faculty that students may have a difficult time envisioning this care in engineering education [66].

We also found that HC awareness paralleled a lot of the emotional responses from participants and among the HC statements and as seen by the regression analysis results. It is possible that the design and presentation of the survey questions may have elicited unconscious or meta-conscious processes that influenced their awareness [24,25] to some of the HC statements or that it elicited indirect interpretations from the participants [23–25,57]. Further work is needed to understand these potential relationships better.

For self-efficacy, most faculties reported high levels of confidence in their ability to succeed despite statements like HC #4 (*"Not everyone can be an engineer"*) suggesting an individuals' internal motivations to persist and belong in such fields (Bonner et al., 2009). However, low levels of self-efficacy were reported for more systemic factors like HC #1 (*"Senior faculty in engineering (e.g., tenured professor) deserve higher status, voice, and have more influence that engineering junior faculty*). This parallels to what the higher education literature states about the status and power faculty gain with the promotion and tenure process [38–40]. For self-advocacy, similar results were found compared to the self-efficacy results suggesting that these two factors may be related although additional participants and work would be needed to confirm these relationships in more detail.

In terms of prevalence, gender and institutional types appeared to have the greatest influences among research-intensive institution (Tier 1). It was interesting to find that among the prevalence of responses, Tier 1 (Ph.D. granting) and Tier 4 (M.S. granting) were more closely similar in influence compared to the other tiers. This is a unique finding and one that will warrant further exploration. It will be necessary to include higher representations of other institutions to identify if this trend still stands if other institutional types are represented in higher numbers.

One interesting observation was that neither faculty rank nor race appeared to result in differences among the faculty responses. For faculty rank, it appears that all faculty regardless of their role (e.g., lecturers, associate professors) are privy to the hidden curriculum of their fields in engineering and that these appear to not change regardless of the discipline of engineering that they may be teaching. Regarding race, it is still unclear if race did not play a role in faculty responses or if these perspectives may have been muffled through the dominant perspectives from the White faculty in this study. Again, added participants to this study can help us clarify these in more detail.

Finally, the three hypotheses were partially confirmed in that all faculty were able to recognize some HC in their fields but our findings also suggested that identification of such were highly contextual (institutional type) and gendered but irrelevant to faculty rank (e.g., lecturers, associate professors). The literature points to the differential experiences that women and intersectional women have compared to their male counterparts [67,68] in engineering [59,69]. However, more studies are needed to understand the influences the formalization of institutional types had on the HC gained from these women faculty in engineering.

5.2. Qual Findings

From the qualitative findings and in response to RQ1, we were able to identify in more detail the lived realities of many of the engineering faculty who responded to this survey. In these, we also confirmed the predominant viewpoints of the inequalities existing for women and intersectional women in engineering in terms of fulfilling the professional expectations from students, peers, and themselves [59,67–69].

While there were many expectations expressed by the engineering faculty, many of them mirrored those found across institutions of higher education [38–40]. However, other field-specific nuances were identified. For example, many engineering faculties indicated attaining accreditation (e.g., ABET) was a characterizing factor of their success as educators. ABET is the leading accreditation that dictates the expectations of the education and training of engineering as is the main agency that faculty and administrators respond to in order to meet the increasing demands of the field [69]. Additionally, the need for rigor in the engineering education curriculum and the level of skill and preparedness of the student in the classroom was also referred to as important by the faculty. This finding parallels what engineering education researchers have referred to as the meritocratic nature of the discipline [44–50,69]. As Stevens and colleagues [69] have suggested, "one of the most significant implications of the meritocracy of difficulty in engineering is how it led engineering students to distinguish themselves from students in other majors and to place their discipline in a clearly superior position to others." ([69], p. 1). It appears that some engineering faculty are mirroring this belief. Additional work is needed to explore this phenomenon further.

One encouraging finding was in the recognition from some faculty, across institutional types and faculty ranks, to advocate and encourage others for changing the status quo of engineering education. It appears that there is an overall sentiment that the system in engineering education has not changed in over a century (e.g., Respondent 32, Associate Professor, Tier 5, Male, White) and that personal and classroom changes through their roles as instructors (e.g., Respondent 34, Assistant Professor, Tier 5, Male, White) can help effect this change. Self-advocacy or a person's willingness to take action and speak up about a matter to improve their quality of life [35] cannot occur without individuals having the motivation or desire to take control over their immediate situation or environment. It appears that for some engineering faculty, this sense of advocacy is becoming an integral component of their professional roles and responsibilities.

6. Limitations

This special issue study was conducted on 55 engineering faculty across different institutions of higher education but the perspectives of multiple faculties within the same institution were not explored. This would have allowed for a more granular study to explore potential similarities or differences between faculties within similar professional environments. Furthermore, more questions of the professionalization of engineering disciplines (e.g., mechanical engineering versus biological engineering) could have provided additional domain- and context-driven of the HC statements and perspectives analyzed in this work.

While it is not the belief that the findings are generalizable as HC is dependent on the individual, the situational, and the contextual [1–8], the findings of this work can help readers to understand how faculty are, or were, influenced by HC. Even though the results of this work may not represent the findings that were attained from the fully validated instrument, they represent the initial interpretations of faculty to our questions during the process of validation. Future studies will compare and contrast the findings from this study on engineering faculties and other participant groups (e.g., graduate students, undergraduate students, etc.) when data from the fully validated survey is analyzed [21].

Regardless, the findings from this special issue study can begin to shed light into the perspectives and mechanism by which engineering faculty understand HC, which may help scholars to see and understand the lived realities of many of these professionals.

7. Conclusions and Recommendations

The findings of this special issue study suggest a possible influence gender and institutional type in the beliefs, values, and attitudes that faculties carry about the HC in engineering. We also found that while some faculties are interested or are using their current roles to advocate on issues of engineering to their students and peers. However, some of the challenges for this action appear to relate to issues of gender and intersectionality, as well as race and institutional type.

From this work, the authors encourage the readers and, in particular, researchers and administrators, to conduct similar types of hidden curriculum studies internally at their institutions and colleges of engineering to explore the dynamics of the professional environments that different faculties are a part of. The authors call for a closer examination to the degree by which institutional resources are responding to the intersectional experiences of faculty (e.g., unequitable workloads for minoritized faculty). Finally, the authors encourage that for all faculty (tenure and non-tenure track), career trajectories and promotional paths become clearer to ensure transparency in evaluation processes.

8. Implications

This study has several implications. The first is that through a mixed-method, hidden curriculum approach, more awareness on the professional needs of each institutional type can be elevated. Second, the need for more individually- and culturally-responsive resources and interventions in engineering are presented. The findings of this special issue study purposed readers to reflect upon the connections that mechanisms such as emotions, self-efficacy, and self-advocacy can play in empowering and enabling all members of an academic engineering faculty group to participate in an equitable and safe environment. Finally, the techniques used in this work (i.e., decision tree) introduces a new way to handle complex datasets such as these and use them to more deeply inform the research and educational communities on the influences that professionalization of engineering faculty can have in their experiences at their colleges and institutions.

Author Contributions: Project Administration, Resources, Supervision, Conceptualization, Design, Methodology, Writing-Original Draft, Writing-Review and Editing: I.V.; Qualitative data analysis and Writing-Original Draft: T.C.; Qualitative Data Collection and Analysis and Writing-Review and Editing: M.D.S.; Quantitative Data Analysis and Writing-Review and Editing: M.T.H.K.

Funding: This material is based upon work supported by the National Science Foundation (NSF) No. EEC-1653140. Any opinions, findings, and conclusions or recommendations expressed in this material does not necessarily reflect those of NSF.

Conflicts of Interest: The authors declare no conflict of interest. The funders had no role in the design of the study; in the collection, analyses, or interpretation of data; in the writing of the manuscript; or in the decision to publish the results.

References

1. Giroux, H.A. *Theory and Resistance in Education: Towards a Pedagogy for the Opposition*; Greenwood Publishing Group: Westport, CT, USA, 2001.
2. Kentli, F.D. Comparison of hidden curriculum theories. *Eur. J. Educ. Stud.* **2009**, *1*, 83–88.
3. Nieto, S. *Affirming Diversity: The Sociopolitical Context of Multicultural Education*; Longman: White Plains, NY, USA, 1992; ISBN 13 978-0131367340.
4. Villanueva, I.; Gelles, L.; Di Stefano, M.; Smith, B.; Tull, R.; Lord, S.; Benson, L.; Hunt, A.; Riley, D. What does hidden curriculum in engineering look like ad how can it be explored? In Proceedings of the American Society of Engineering Education Annual Conference and Exposition, Minorities in Engineering Division, Salt Lake City, UT, USA, 24–27 June 2018; pp. 16.
5. Erickson, S.K. Engineering the Hidden Curriculum: How Women Doctoral Students in Engineering Navigate Belonging. Ph.D. Thesis, Arizona State University, Tempe, AZ, USA, 2007.
6. Tormey, R.; Le Duc, I.; Isaac, S.; Hardebolle, C.; Cardia, I. The formal and hidden curricula of ethics in engineering education. In Proceedings of the 43rd Annual SEFI Conference, Orléans, France, 29 June–2 July 2015.
7. Haefferty, F.W.; Gaufberg, E.H. The hidden curriculum. In *A Practical Guide for Medical Teachers*, 5th ed.; Dent, J.A., Harden, R.M., Hunt, D., Eds.; Elsiever: Beijing, China, 2017; pp. 35–43.
8. Rabah, I. The influences of assessment in constructing a hidden curriculum in higher education: Can self and peer assessment bridge the gap between the formal and hidden curriculum? *Int. J. Hum. Soc. Sci.* **2012**, *2*, 236–242.
9. Borges, J.C.; Ferreira, T.C.; de Oliveira, M.S.B.; Macini, N.; Caldana, A.C.F. Hidden curriculum in student organizations: Learning, practice, socialization, and responsible management in a business school. *Int. J. Manag. Educ.* **2017**, *15*, 153–161. [CrossRef]
10. Baird, J.; Bracken, K.; Grierson, L.E. The relationship between perceived preceptor power use and student empowerment during clerkship rotations: A study of hidden curriculum. *Med. Educ.* **2016**, *50*, 778–785. [CrossRef] [PubMed]
11. Cotton, D.; Winter, J.; Bailey, I. Researching the hidden curriculum: Intentional and unintended messages. *J. Geogr. Higher Educ.* **2013**, *37*, 192–203. [CrossRef]
12. Joughin, G. The hidden curriculum revisited: A critical review of research into the influence of summative assessment on learning. *Assess. Eval. Higher Educ.* **2010**, *35*, 335–345. [CrossRef]
13. Margolis, E. *The Hidden Curriculum in Higher Education*; Routledge: New York, NY, USA, 2001; ISBN 0-415-92758-7.
14. Smith, B. *Mentoring at-Risk Students through the Hidden Curriculum of Higher Education*; Lexington, Books: Plymouth, UK, 2014; ISBN-13: 978-1498515801.
15. Lipsitt, D.R. Developmental life of the medical student: Curriculum considerations. *Acad. Psychiatry* **2015**, *39*, 63–69. [CrossRef] [PubMed]
16. Riva, S.; Monti, M.; Ianello, P.; Pravettoni, G.; Schulz, P.J.; Antonietti, A. A preliminary mixed-method investigation of trust and hidden signals in medical consultations. *PLoS ONE* **2014**, *9*, e90941. [CrossRef] [PubMed]
17. Street, R.L., Jr.; Gordon, H.; Haidet, P. Physicians' communication and perceptions of patients: Is it how they look, how they talk, or is it just the doctor? *Soc. Sci. Med.* **2007**, *65*, 586–598. [CrossRef] [PubMed]
18. Young, M.E.; Norman, G.R.; Humphreys, K.R. The role of medical language in changing public perceptions of illness. *PLoS ONE* **2008**, *3*, e3875. [CrossRef] [PubMed]
19. Mead, R.; Hejmadi, M.; Hurst, L.D. Teaching genetics prior to teaching evolution improves evolution understanding but not acceptance. *PLoS Biol.* **2017**, *15*, e2002255. [CrossRef] [PubMed]
20. Alsubaie, M.A. Hidden Curriculum as One of Current Issue of Curriculum. *J. Educ. Pract.* **2015**, *6*, 125–128.
21. Villanueva, I.; Di Stefano, M.; Gelles, L.; Youmans, K.; Hunt, A. Development and validation of a mixed-methods vignette survey to explore hidden curriculum in engineering. *JEE*, under review.
22. Hargreaves, A. *Teaching in the Knowledge Society: Education in the Age of Insecurity*; Teachers College Press: New York, NY, USA, 2003.
23. Winkielman, P.; Schooler, J.W. Splitting consciousness: Unconscious, conscious, and metaconscious processes in social cognition. *Eur. Rev. Soc. Psychol.* **2011**, *22*, 1–35. [CrossRef]

24. Winkielman, P.; Berridge, K.; Sher, S. Emotion, consciousness, and social behaviour. In *Handbook of Social Neuroscience*; Decety, J., Cacioppo, J.T., Eds.; Oxford University Press: Oxford, UK, 2011; pp. 195–211.

25. Zeeman, A. *Consciousness: A User's Guide*; Yale University Press: New Haven, CT, USA, 2002.

26. Koriat, A. *Metacognition and Consciousness*; Institute of Information Processing and Decision Making, University of Haifa: Haifa, Israel, 2006.

27. Schooler, J.W. Discovering memories in the light of meta-awareness. *J. Aggress. Maltreat. Trauma* **2001**, *4*, 105–136. [CrossRef]

28. Schooler, J.W. Re-representing consciousness: Dissociation between experience and meta-consciousness. *Trends Cogn. Sci.* **2002**, *6*, 339–344. [CrossRef]

29. Pekrun, R.; Linnenbrink-Garcia, L. Emotions in education: Conclusions and future directions. In *International Handbook of Emotions in Education*, 1st ed.; Pekrun, R., Linnenbrink-Garcia, L., Eds.; Taylor Francis: New York, NY, USA, 2014; pp. 659–675. ISBN 9780415895019.

30. Bandura, A. Perceived self-efficacy in cognitive development and functioning. *Educ. Psychol.* **1993**, *28*, 117–148. [CrossRef]

31. Bandura, A. *Self-Efficacy: The Exercise of Control*, 1st ed.; Worth Publishers: London, UK, 1997; SBN-13: 978-0716728504.

32. Caprara, G.V.; Di Giunta, L.; Eisenberg, N.; Gerbino, M.; Pastorelli, C.; Tramontano, C. Assessing Regulatory Emotional Self-Efficacy in Three Countries. *Psychol. Assess.* **2008**, *20*, 227–237. [CrossRef] [PubMed]

33. Sugarman, J.; Sokol, B. Human agency and development: An introduction and theoretical sketch. *New Ideas Psychol.* **2012**, *30*, 1–14. [CrossRef]

34. Schreiner, M. Effective self-advocacy: What students and special educators need to know. *Interv. Sch. Clin.* **2007**, *42*, 300–304. [CrossRef]

35. Ryan, T.G.; Griffits, S. Self-advocacy and its impacts for adults with developmental disabilities. *Aust. J. Adult Learn.* **2015**, *55*, 31–55.

36. Ozolins, I.; Hall, H.; Peterson, R. The student voice: Recognising the hidden and informal curriculum in medicine. *Med. Teach.* **2008**, *30*, 606–611. [CrossRef] [PubMed]

37. Hundert, E.M.; Hafferty, F.; Christakis, D. Characteristics of the informal curriculum and trainee's ethical choices. *Acad. Med.* **1996**, *71*, 624–630. [CrossRef] [PubMed]

38. Gerber, J. Professionalization as the basis for academic freedom and faculty governance. *AAUP J. Acad. Freedom* **2010**, *1*, 1–26.

39. Abbott, A. *The System of Professions: An Essay on the Division of Expert Labor*; The University of Chicago Press: Chicago, IL, USA, 1988.

40. Wilensky, H.C. The professionalization for everyone. *Am. J. Sociol.* **1964**, *7*, 137–158. [CrossRef]

41. Larson, M.S. *The Rise of Professionalism: A Sociological Analysis*; University of California Press: Los Angeles, CA, USA, 2017.

42. Ross, D. *The Origins of American Social Science*; Cambridge University Press: New York, NY, USA, 1991.

43. Brubacher, J.S.; Rudy, W. *Higher Education in Transition*, 3rd ed.; Harper & Row: New York, NY, USA, 1976.

44. Villanueva, I.; Nadelson, L. Are we preparing our students to become engineers of the future or the past? *Int. J. Eng. Educ.* **2017**, *33*, 639–652.

45. Davis, M. Engineering as profession: Some methodological problems in its study. In *Engineering Identities, Epistemologies, and Values*, 1st ed.; Christensen, S.H., Didier, C., Jamison, A., Meganck, M., Mitcham, C., Newberry, B., Eds.; Springer: Cham, Switzerland, 2015; pp. 65–98. ISBN 978-3-319-16172-3.

46. Downey, G.L.; Lucena, J.C. Knowledge and professional identity in engineering: Code-switching and the metrics of progress. *Hist. Technol.* **2004**, *20*, 393–420. [CrossRef]

47. Baillie, C.; Pawley, A.L.; Riley, D. (Eds.) *Engineering and Social Justice*; Purdue University Press: West Lafayette, IN, USA, 2012.

48. Jolly, L.; Radcliffe, D. Reflexivity and hegemony: Changing engineers. In Proceedings of the 23th HERDSA Annual Conference, Toowoomba, Australia, 2–5 July 2000; pp. 357–365.

49. Riley, D. Engineering and social justice. In *Synthesis Lectures on Engineers, Technology, and Society*; Morgan and Claypool: San Rafael, CA, USA, 2008; Volume 3, pp. 1–152. [CrossRef]

50. Lohmann, J.R.; Froyd, J.E. Chronological and ontological development of engineering education as a field of scientific inquiry. In *Cambridge Handbook of Engineering Education Research*; Johri, A., Olds, B.M., Eds.; Cambridge University Press: Cambridge, MA, USA, 2010; pp. 1–25, ISBN-13: 978-1107014107.

51. Creswell, J.W. *Research Design*, 3rd ed.; Sage Publications, Inc.: Thousand Oak, CA, USA, 2008.
52. Creswell, J.W.; Plano Clark, V.L. *Designing and Conducting Mixed Methods Research*, 3rd ed.; Sage publications: Thousand Oaks, CA, USA, 2018.
53. Bryman, A. Integrating quantitative and qualitative research: How is it done? *Qual. Res.* **2006**, *6*, 97–113. [CrossRef]
54. Carnegie Classification of Institutions of Higher Education. 2018. Available online: http:// carnegieclassifications.iu.edu/classification_descriptions/basic.php (accessed on 17 September 2018).
55. Loftland, J.; Snow, D.; Anderson, L.; Loftland, L.H. *Analyzing Social Settings: A Guide to Qualitative Observation and Analysis*; Wadsworth, Cengage Learning: Belmont, CA, USA, 2006.
56. Saldaña, J. *The Coding Manual for Qualitative Researchers*, 3rd ed.; SAGE Publications: Los Angeles, CA, USA, 2016.
57. Horn, R.A. Developing a critical awareness of the hidden curriculum through media literacy. *JSTOR* **2003**, *76*, 298–300. [CrossRef]
58. Gilliam, F.D., Jr.; Bales, S.N. Strategic Framing Analysis: Reframing America's Youth. In *Social Policy Report, Giving Child and Youth Development Knowledge Away*; Society for Research in Child Development: Washington, DC, USA, 2001; Volume 15, Available online: http://www.srcd.org/sites/default/files/documents/spr15-3.pdf (accessed on 20 August 2018).
59. DeCuir-Gunby, J.T.; Long-Mitchell, L.A.; Grant, C. The emotionality of women professors of color in engineering: A critical race theory and critical race feminism perspective. In *Advances in Teacher Emotion Research*; Springer: Boston, MA, USA, 2009; pp. 323–342.
60. Alm, C.O.; Roth, D.; Sproat, R. Emotions from text: Machine learning for text-based emotion prediction. In Proceedings of the Conference on Human Language Technology and Empirical Methods in Natural Language Processing, Vancouver, BC, Canada, 6–8 October 2005; Association for Computational Linguistics: Stroudsburg, PA, USA, 2005; pp. 579–586.
61. Field, A. *Discovering Statistics Using IBM SPSS*, 4th ed.; SAGE: London, UK, 2013.
62. Song, Y.; Lu, Y. Decision tree methods: Applications for classification and prediction. *Shanghai Arch. Psychiatry* **2015**, *27*, 130–135. [CrossRef] [PubMed]
63. Khan, M.T.H.; Ahmed, F.; Kim, K.Y. Integration and visualization framework for data-driven resistance spot welded assembly design. In Proceedings of the 2017 13th IEEE Conference on Automation Science and Engineering (CASE), Xi'an, China, 20–23 August 2017.
64. Khan, M.T.H.; Demoly, F.; Kim, K.Y. Dynamic design intents capture with formal ontology and perdurants object concept for collaborative product design. In Proceedings of the Collaboration Technologies and Systems (CTS) International Conference, Orlando, FL, USA, 31 October–4 November 2016.
65. ABET Board of Directors. *Criteria for Accrediting Engineering Programs: Effective for Reviews during the 2018–2019 Accreditation Cycle*; Engineering Accreditation Commission: Baltimore, MD, USA, 2018; pp. 1–45.
66. Capobianco, B.M.; Yu, J.H. Using the construct of care to frame engineering as a caring profession toward promoting young girls' participation. *J. Women Minor. Sci. Eng.* **2014**, *20*, 21–33. [CrossRef]
67. Crenshaw, K. Mapping the margins: Intersectionality, identity politics, and violence against women of color. *Stanf. Law Rev.* **1991**, 1241–1299. [CrossRef]
68. Davis, K. Intersectionality as buzzword: A sociology of science perspective on what makes a feminist theory successful. *Fem. Theory* **2008**, *9*, 67–85. [CrossRef]
69. Stevens, R.; Amos, D.M.; Garrison, L.; Jocuns, A. Engineering as lifestyle and a meritocracy of difficulty: Two pervasive beliefs among engineering student and their possible effects. In Proceedings of the American Society for Engineering Education Annual Conference, Honolulu, HI, USA, 24–27 June 2007; pp. 1–2.

education sciences

MDPI

Article

Engineering Projects in Community Service (EPICS) in High Schools: Subtle but Potentially Important Student Gains Detected from Human-Centered Curriculum Design

Alissa Ruth [1,*], Joseph Hackman [1], Alexandra Brewis [1], Tameka Spence [1], Rachel Luchmun [1], Jennifer Velez [2] and Tirupalavanam G. Ganesh [2]

[1]   School of Human Evolution and Social Change, Arizona State University, Tempe, AZ 85281, USA;
      joseph.hackman@asu.edu (J.H.); Alex.brewis@asu.edu (A.B.); Tameka.spence@asu.edu (T.S.);
      rluchmun@asu.edu (R.L.)
[2]   Ira A. Fulton Schools of Engineering, Arizona State University, Tempe, AZ 85281, USA;
      Jennifer.Velez@asu.edu (J.V.); tganesh@asu.edu (T.G.G.)
*   Correspondence: alissa.ruth@asu.edu

Received: 31 December 2018; Accepted: 30 January 2019; Published: 7 February 2019

Abstract: A major goal in Engineering training in the U.S. is to continue to both grow and diversify the field. Project- and service-based forms of experiential, problem-based learning are often implemented with this as a goal, and Engineering Projects in Community Service (EPICS) High is one of the more well-regarded and widely implemented. Yet, the evidence based on if and how participation in such programs shapes student intentions and commitment to STEM pathways is currently limited, most especially for pre-college programming. This study asks: How do high school students' engineering mindsets and their views of engineering/engineers change as they participate in project–service learning (as implemented through an EPICS High curriculum)? This study employed a mixed method design, combining pre- and post-test survey data that were collected from 259 matched students (63% minority, 43% women) enrolling in EPICS High (total of 536 completed pre-tests, 375 completed post-tests) alongside systematic ethnographic analysis of participant observation data conducted in the same 13 socioeconomically diverse schools over a two-year period. Statistical analyses showed that participants score highly on engineering-related concepts and attitudes at both pre- and post-test. These did not change significantly as a result of participation. However, we detected nuanced but potentially important changes in student perspectives and meaning, such as shifting perceptions of engineering and gaining key transversal skills. The value of participation to participants was connected to changes in the meaning of commitments to pursue engineering/STEM.

Keywords: high school; engineering curriculum; STEM; service-learning; project-based learning; underrepresented minorities; outcomes

1. Introduction

Professional Engineering training is not currently meeting perceived national needs for competitiveness, meaning there is a push to recruit, and then retain to graduation, larger cohorts of suitable students [1]. In addition, even though relative gains have been made, the goal of equitable representation of women and historically-underrepresented minority (URM) students in the Science, Technology, Engineering, and Math (STEM) fields in the U.S. has also not yet been achieved. For example, even though women now earn the majority of bachelor's degrees overall, they represent only 20% of those awarded in Engineering. Similarly, African-American students graduate with 9.5% of the undergraduate degrees in the U.S. but represent only 3.8% of those in Engineering [2].

This underrepresentation in STEM begins prior to university entrance, being clearly evident in high school. For example, Asian and White students are much more likely to successfully complete calculus by the 12th grade compared to URM students [3]. Thus, there has been a significant national push to find the best means to both engage and build the relevant skills for diverse students, and students more generally, as potential STEM graduates while they are still in the K-12 system. A fundamental goal of this broader agenda is to integrate pedagogies that focus on collaborative forms of community service, practical application, and project-based learning that articulate solutions for real-world problems. Innovative high school mathematics and science curricula have been developed along these lines to address this challenge, both aiming to orient students towards and prepare them better for advanced study in these fields [4,5]. One of the most widely applied is EPICS (Engineering Projects in Community Service) High [6].

EPICS is one of the more widely applied means to meet this goal. The EPICS strategy is based on connecting students interested in engineering and computing design with local community partners to solve practical problems [7,8], including over extended semesters [9]. This strategy of "designing for others" [10] was originally intended for engaging college students, with the assumption that project-based and service-learning modalities would spark a broader interest in and commitment to STEM careers. The underpinning philosophy for Engineering, for example, is that highlighting and engaging Engineering's concern for people, local communities, and broader societal welfare would better draw women and URM students into the field [9]. The approach of EPICS is consistent with an array of published research that indicates that use of dynamic curricula that incorporate service-learning and/or project-based learning predicts improved student outcomes in potentially advancing minority and female students in engineering/STEM at the college level [7,11–15].

In 2006, responding to this recognition that attrition from the STEM training pipeline was occurring earlier in students' education and needed to be addressed sooner, the program developed a high school curriculum [16]. This fits with research showing that students who excel in mathematics in high school (specifically 10th- through 12th-grade math) and have self-belief in their mathematic abilities are more likely to pursue STEM degrees in college [14]. Early exposure to mathematics and science related courses seems to be key but should be done in a way that *engages* students and makes learning these subjects enjoyable [17–19]. Project/problem-based learning, particularly in early grades, can act to foster creativity through problem-solving, a much-needed skill in engineering [18]. Thus, earlier education is considered central to instilling the necessary skills and knowledge to create successful STEM pathways for future practitioners [19].

The EPICS curriculum was subsequently deployed in high schools across the country ("EPICS High"). At the high school level, the technical capacities of students to engage in human-centered engineering solutions to real work may be less [12], but the underlying assumptions remains the same: That such programs should: (a) Fundamentally change students' perceptions of engineering by making the engineering design process fully human-centered and (b) that the greatest positive shifts in perception should be for URM and female students.

Operationally, EPICS High programming is integrated into existing classrooms and school arrangements. During the summer, math/science teachers are trained on the EPICS High curriculum, which is then incorporated into their STEM or career and technical education (CTE) classes, or in afterschool clubs. Usually, projects seek to engage with the needs of partners local to the specific school. The EPICS High approach fits within a broader trend in STEM education toward supporting innovations in STEM-project-based learning (PBL) [12,14,20]. Unlike *problem*-based learning—where students are given a hypothetical, real-world problem to solve through thinking through different steps [20]—*project*-based learning includes actual projects that have clients and real applications [11,12,14]. Specifically, PBL mimics professional work in that the projects take longer to complete, it applies knowledge rather than simply acquiring it, and students must self-direct their time and energies as they work on a team [14]. In general, student outcomes from participation in PBL include improved teamwork, greater communication skills, a better understanding of the complexity of professional problem-solving and the knowledge needed in order to apply solutions, as well as increased motivation to continue on [11,12]. Conversely, because of the project–solution focus, students may be deficient in understanding the fundamentals of engineering concepts [12]. Much of the research suggests the gains are seen in all participants, not just URM and female students.

The EPICS High approach also purposefully focuses student activity in service, not just project learning. As a form of pedagogy, service-learning is experiential education that educators design as structured opportunities where students partake in activities that address community needs in order to foster learning and development [21] (p. 5). Within engineering, service-learning can help students to build skills useful to the prolonged professional practice of engineering while also fostering civic responsibilities [15]. Service-learning is quite distinctive because, when done most effectively, it gives students a voice to choose what project they want to complete and it involves community partners—both factors that help to motivate students to gain the best learning outcomes as well as make the experience meaningful [22,23]. For example, students who fully engage in service-learning report enjoying school more and becoming more civically inclined [23]. Furthermore, students who engage in service-learning that incorporates STEM problem-based learning techniques increase their academic engagement, increase achievement in science, and become more resilient and civically engaged [13].

The *college-based* Engineering Projects in Community Service (EPICS) program typically integrates both service-learning and project-based learning into one program that spans one or more semesters. College students' reflections on EPICS programs suggest students believe they gained skills in teamwork, communication, project planning, leadership, and that engineering could be viewed as a "caring profession" [6,9,24]. Moreover, Purdue University enrollment data over a decade suggest female students in Mechanical and Electrical and Computer Engineering majors were 170% more likely than male students to participate [9,25]. Using a case-study approach, Huff et al. (2012) [24] suggested benefits in this engaged learning (in international settings) also had value for advancing basic engineering competencies.

EPICS-related gains at high school levels are even less well studied, but early signs suggest that impacts might be hampered by a lack of basic skills (such as mathematics). In a pilot study comparing EPICS High program participants to those in another program without the service-learning components, Kelly et al. (2010) [26] reported participants in both programs exhibited significant challenges in solving basic problems; differences by gender or ethnicity were not studied (the sample was small) [26]. Zoltowski, Oakes, and Cardella (2012) [27] used a qualitative–phenomenological approach to identify the ways that 33 students, enrolled in EPICS and in several similar programs, understood and reacted to the idea of engineering as a human-centered design. They concluded that students understood the needs of the end-user and were able to integrate those needs into their designs. Although the sample was selected for diversity, the role of this diversity in shaping perceptions was not an interpretive focus.

Considering the broader potential of EPICS and similar instructional modalities for changing the engineering pipeline at the high school level, the evidence base must include a systematic assessment of how these students might be differently impacted; many of the available studies are preliminary in nature. Based on the demographics of students participating in 50 schools in three states, Oakes et al. (2012) [28] concluded relative participation by girls (44%) and URM students and a high percentage of students eligible for free or reduced school lunches (46%) were markers of success. In addition, of the students who said at the beginning of the program they would "not at all" be interested in a STEM major, 53% of girls and 47% of boys had subsequently changed their response to "a lot".

To our knowledge, however, there are currently no detailed studies focused on how EPICS High (or similarly designed project- and service-learning curricula) might differently impact women and minorities compared to other students—i.e., those that the fields of Engineering and STEM more generally are seeking to advance into careers. Our goal in this study was to do just that, as a first step in establishing a solid evidence base for identifying which strategies should best help to meet the much larger goal of supporting and diversifying the student pipeline in Engineering and related STEM fields. Our research question is: How do high school students' engineering mindsets and their views of engineering/engineers change as they participate in project-service learning (as implemented through an EPICS High curriculum)?

To answer this, we used pre- and post-test data to identify if the degree of change after program completion was predicted by low beginning scores. Since the theory behind EPICS High curricula is of value to enhancing a diverse STEM pipeline because project–service-based learning changes students perceptions, our research strategy engaged in collecting and analyzing qualitative data on student perceptions alongside quantitative ratings of pre- and post-test mindsets.

2. Materials and Methods

Arizona State University (ASU) began delivering the EPICS college program in 2009 and then collaborated to deliver an EPICS High curriculum with local high schools that could act as potential pipelines to their Engineering degrees. Now the ASU program serves as one of three EPICS High hubs, meaning that EPICS High schools in the large metropolitan area work directly with our university program, whereas other EPICS High programs work with Purdue University. As the largest hub, the ASU program currently serves just over 800 high school students within 32 schools in the Phoenix Metro area, and the EPICS programming specifically makes it a point to partner with more diverse student bodies. For example, out of the 32 schools, 13 are classified as serving low-income, high ethnic-minority schools. The core curriculum mirrors Purdue's EPICS High (curriculum can be found at https://engineering.purdue.edu/EPICS/k12), defined by student engagement in the design process (Figure 1). EPICS High at ASU is a highly coordinated endeavor with dedicated staff to implement the program and support the high school teachers throughout the year. Teachers are given access to the curriculum via Purdue University's website portal, provided a week-long summer training, have dedicated college student mentors who visit the classrooms regularly, are offered a funding competition to support project development throughout the school year, and host a showcase as a culminating experience at the end of the spring semester. The EPICS High model is integrated into existing classroom frameworks, either through their STEM or career and technical education (CTE) classes or in afterschool clubs. The curriculum is grounded in design education and service-learning pedagogies and seeks to promote engineering as a force for social good.

EPICS Design Process

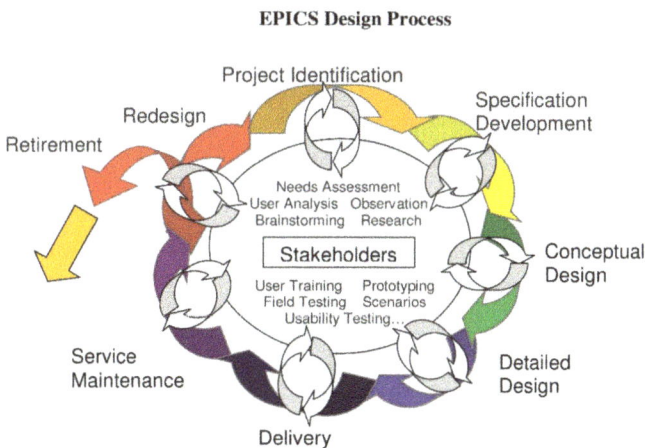

Figure 1. The engineering projects in community service (EPICS) design process (reproduced from EPICS website).

By pairing meaningful community service with engineering instruction, EPICS High seeks to provide a conduit for students to engage in project-based learning to master course content while fostering greater civic responsibility and community engagement. Moreover, the curriculum incorporates human-centered design—focusing on the needs and uses for the end-stakeholder—and key engineering processes to foster engineering habits of mind, such as systems thinking, optimism, and ethical consideration in engineering, as well as entrepreneurial mindsets such as the three Cs (curiosity, connections, and creation of value) from the KEEN Framework [29]. Across EPICS High programming, students continuously explore, at increasing levels of sophistication, solutions to problems identified by their community partners by applying skills they are learning in the classroom. Ultimately, students work with members of the community to create engineering solutions to address real-world problems (see Table 1 for selection of projects).

Table 1. Sample selection of student team projects. STEM: Science, Technology, Engineering, and Math.

Community Partner	Project Goals
Audubon Society	Redesign and renovate the seating and shading structure of the Butterfly Garden, a space used to host community environmental events and school field trips
County Animal Care and Control	Create a durable, hygienic, and inexpensive dog bed for animal shelters
Student's Own High School	Rebuild their school hypnotherapy garden to make it more accessible for those with physical disabilities
Local STEM Outreach Program	Teach children in foster homes about STEM to create better opportunities
Neighboring Elementary School	Create an app that is tailored to the curriculum to enhance student reading skills
Horse Rescue	Create a water catchment system to avoid stalls becoming flooded when it rains
Family-centered not-for-profit organization	Improve soundproofing of the community room in a low-income family housing complex
Students' Own High School Community	Build a community garden to provide fresh fruits and vegetables to members of the community (the high school is located in a food desert)
Homeless Youth Organization	Assess the location most in need of help for the organization to expand
Housing/Health/Community Service Organization	*Design a robotics curriculum for a summer program for homeless children aged 5 to 12*

Data for this study were collected during the 2016–2017 and the 2017–2018 school years, covering two separate sequential implementations of the EPICS High program. Human subjects' approval for this study was acquired through the ASU Institutional Review Board (STUDY00004523), and each school also provided individual administrative approval.

2.1. Scalar Data Collection and Analysis

We began data collection with an online survey targeting all participating students, deployed at the beginning of each school year. Surveys ($N = 838$) were completed by high school students from 15 schools (Wave 1 $N = 361$, Wave 2 $N = 344$). The surveys included a set of items for pre-testing of outcomes (see "scale construction" below). The second wave survey also presented open-ended narrative responses to questions to be assessed through systematic qualitative analysis. These questions included their goals after high school and their views of engineering, pre- and post-test. Post-test surveys were given at the end of each spring semester. We excluded from post-tests any students who proved from the pre-test to be outside of the grade range for the EPICS High program (6th and 7th graders, $N = 10$), as well as all students from the one high school that failed to implement the EPICS program in full during the semester ($N = 121$ students). After these exclusions, we had a total of $N = 705$ unique students across both waves. Unforeseen issues with the unique identifiers (required for ethics protection), and follow-up response rates, resulted in additional students being excluded from the post-study pool. This resulted in a final matched pre/post-test sample of 259 (Table 2). Students self-identified ethnicity and gender as part of the survey demographics.

Table 2. Participant sample size and demographics.

	Pre-test (N = 578)		Post-test (N = 386)		Matched (N = 259)	
Grade	N	%	N	%	N	%
8th	35	6.1	15	3.9	9	3.5
9th	44	7.6	39	10.1	26	10
10th	184	31.8	110	28.5	73	28.2
11th	205	35.5	122	31.6	89	34.4
12th	94	16.3	90	23.3	62	23.9
Missing	16	2.8	10	2.6	0	0
Gender						
Male	312	54	218	56.5	142	54.8
Female	225	38.9	143	37	110	42.5
Prefer not to respond	21	3.6	15	3.9	7	2.7
Missing	20	3.5	10	2.6	0	0
Ethnicity						
White	226	39.1	144	37.3	96	37.1
Latino(a)	200	34.6	137	35.5	98	37.8
Asian	48	8.3	45	11.7	34	13.1
URM (Latinx, African American)	43	7.4	23	6.0	17	6.6
Missing/Refused	61	10.6	37	9.6	14	5.4
Parent College Graduate						
None	345	59.7	196	50.8	126	48.6
One Parent	118	20.4	92	23.8	60	23.2
Both Parents	115	19.9	98	25.4	73	28.2
Parents Engineer						
Parent an engineer	94	16.3	65	16.8	50	19.3
Title I						
Non-Title I School	217	37.5	170	44	134	51.7
Title I School	361	62.5	216	56	125	48.3
Total	578	100	386	100	259	100

The surveys contained 23 items for pre/post-test using scale items directly related to the learning outcomes for EPICS High (Table 3) and adapted from scales developed to assess the KEEN Framework's 3Cs [30,31]. These items were selected to assess status (and thus growth) in the following domains: Attitudes towards engineering (learning about engineering, considering studying engineering, understanding the importance of engineering); improving ideas (inventing new ways of doing things); importance of feedback (identifying needs of stakeholders, seeking input, incorporating feedback into designs); growth mindset (seeing obstacles as opportunities, not giving up on difficult tasks, seeing failure as a chance to improve); social responsibility (contributing to the good of society, seeking opportunities to improve lives of others); and importance of multi-perspectives (putting self in other's shoes, incorporating different expertise/ideas). The item responses were on 5-point Likert scale ranging from strongly agree (5) to strongly disagree (1). These scales demonstrated a high reliability across both the pre- and post-test (Table 3). Students overall scored highly on all scale items, resulting in a left-skewed distribution for all scales at both pre-test and post-test.

Given we were only able to match 259 students across pre-test and post-test surveys, we performed more liberal statistical analyses on the scalar responses for the full analytic sample, comparing all pre-test scores with all post-test scores using the Mann–Whitney U test for binary groups and the Kruskal–Wallis test for tests among multiple groups. We then focused a second set of more conservative analyses on the matched sample and tests for growth within students from pre-test to post-test for those surveyed twice using the Wilcoxon signed-test.

Table 3. Scale reliability and the original contributing scale items.

	Pre-test Scale Reliability		Post-test Scale Reliability	
	N	Alpha	N	Alpha
Improve Ideas	562	0.79	375	0.83
Importance of Feedback	563	0.83	370	0.86
Growth Mindset	557	0.73	372	0.71
Social Responsibility	560	0.82	371	0.85
Importance of Multiple Perspectives	562	0.79	372	0.82
Attitudes towards Engineering	559	0.72	375	0.63

2.2. Qualitative Data Collection and Analysis

Qualitative data collection, in the form of open-ended questions to students, was applied via two different methods. First, in the wave 2 (pre- and post-test) surveys, we added open-ended elicitations, asking students to reflect on the program. Examples of the questions were: "What are your views of engineers?" and "What are your plans after high school?" Second, over the two years of the study, we conducted extended participant observation—overseen by a PhD anthropologist—across the 13 schools. The ethnographic procedures included regular site visits and extended note-taking on informal interviews with students and teachers at least at two points in time for each school, once in the beginning of fall semester and again towards the end of the program in the spring semester. Since our research questions focused on the experiences of students, more detailed ethnographic attention was given to six schools designated as Title I as well as women and ethnic minorities across school types. The Title I label signifies that they serve low-income students, and these schools typically have a large number of ethnic minority students. For example, in Arizona, 41.5% of students in Title I schools are White, and Latinos represent 43% of the school population [32]. The site visits included classroom observations, documenting the forms and levels of student engagement with the curriculum, and talking informally with as many students as possible as many times as possible about their curricular experiences and evolving community projects. These informal interviews are unstructured and conversational in form, and the procedure is for the interviewer to take copious notes during the conversation and/or immediately afterward, including quotes [33]. The trained ethnographic researcher conducting these informal interviews was female, non-White, and relatively young; it was hoped this would support greater disclosure by the students. The site visits also involved taking copious field notes on the following: Student motivations for joining EPICS High; motivations for enrolling in engineering; students' experiences working and designing for a stakeholder; and student team dynamics, among other topics. In these interactions, some students offered direct or indirect information on their ethnicity, but the researcher did not directly ask. Findings from the data set should be assessed with this in mind, and this also explains why we more simply coded students generally as White or non-White solely as indicative and did not take a further step of conducting URM versus other student comparisons.

Using systematic methods of qualitative data coding and analysis [34], we analyzed the resulting detailed field notes in addition to the narrative responses (Section 2.2, N = 215) of pre/post-test surveys. This process yielded a total of 60,324 words of text. Then again, using normal procedures for code generation and assignment, we developed a codebook and used deductive coding techniques to identify theme repetition within the body of text. Literature on the benefits of K-12 STEM-based project-based curricula, and the literature on developing engineering mindsets, were used as the general guide for what to look for first in the process of code identification [34]. An example of a code is: "Failure and Learning," described as: "Students discuss the ways in which project failures, shortcomings, and mistakes, small and large, impact their learning experiences throughout the class." Once we coded all the text, we were able to retrieve the coded segments to verify we indeed had repeating themes and were then able to make generalizations about the student learning outcomes.

Then, in order to illustrate these analytical findings below, we identified specific exemplars (i.e., coded sections of text that clearly met inclusion criteria for that specific code or code set, and thus exemplified the theme that was identified) [33].

3. Results

3.1. Survey Pre- and Post-Test Scale Results

The scale means pre- and post-test for the full sample are presented in Table 3. The results of the Mann–Whitney U test show that scores at post-test are significantly statistically higher than scores at the pre-test, for all scales (Figures 2 and 3): Improvement of ideas, importance of feedback, growth mindset, social responsibility, importance of multiple perspectives, and attitudes toward engineers/engineering. Assessing the same set of pre-test and post-test scores in the matched sample showed improvement in scores; however, none of the observed differences were statistically significant.

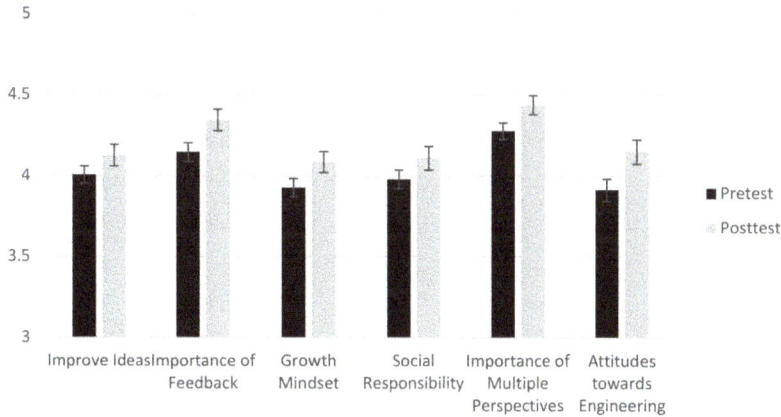

Figure 2. Mean Likert scale response score for pre- and post-test: Full sample.

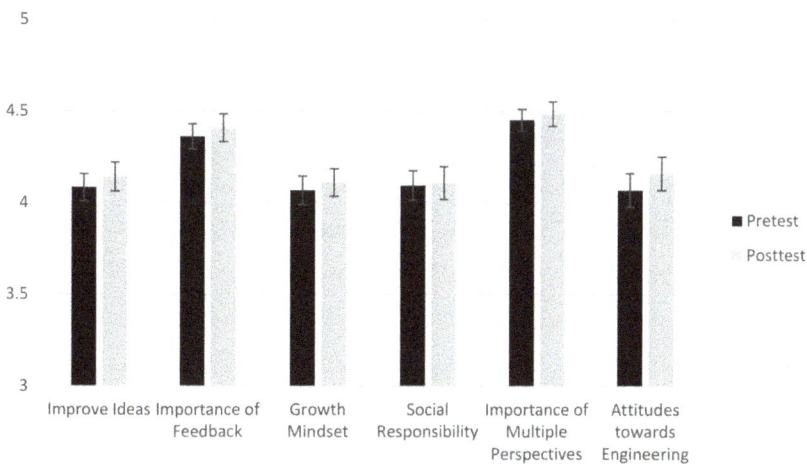

Figure 3. Mean Likert scale response score for pre- and post-test: Matched sample only.

Comparing gender differences in all completed pre-test surveys ($N = 578$) showed female students had significantly lower scores on attitudes toward engineering, improving ideas, importance of feedback, and the importance of multiple perspectives than their male counterparts. However, in all completed post-test surveys ($N = 386$), only differences in attitudes toward engineering remained statistically significant between males and females. In the matched sample analysis ($N = 259$), female students saw significant growth between pre-test scores and post-test scores in the attitudes towards engineering and the importance of multiple perspectives. Additionally, female students had significantly more growth in the importance of feedback compared to males, i.e., the difference between pre- and post-test scores was significantly larger for female students.

In the completed pre-test sample, Title I schools had significantly lower scores on the importance of feedback and the importance of multiple perspectives scales in the full sample. While the scores on these scales increased for both Title I and non-Title I schools, the differences remained significant when assessed in the full post-test sample. Analysis of the matched sample showed none of the differences between pre- and post-test were significant for either Title I or non-Title I schools.

In the full pre-test sample ($N = 578$), the nonparametric ANOVA showed significant differences in the importance of feedback and the importance of multiple perspectives across different ethnicities. Primarily, those self-identified as Asian scored the highest on the importance of feedback and multiple perspectives. Furthermore, these were significantly greater than those in the "other" category who self-identified as African-American, Native American or Pacific Islander, or were missing data on ethnic background. These other and missing categories had the lowest scores on these two scales. Differences in the importance of feedback and multiple perspectives were also observed in the full post-test sample ($N = 386$) and were again driven by the low scores of students in the "other" or missing ethnicity category. In addition, differences emerged at post-test for three of the other scale items: Attitudes toward engineering, improving ideas, and social responsibility. Again, the relatively low scores drove these significant findings for those students in the other or missing ethnicity category. The matched sample ($N = 259$) suggests that these differences were driven primarily by growth among Asian students in their attitudes towards engineering, improving ideas, and social responsibility. These were the only scales to show significant growth, and only among self-identified Asian students.

In the full pre-test sample ($N = 578$), results showed significant differences in the scores for attitudes toward engineering, importance of feedback, and the importance of multiple perspectives across varying levels of parental education. Attitudes toward engineering were similar among students with parents having no college degree, or with one parent having a college degree. However, the differences in the importance of feedback and multiple perspectives were driven by students with parents who did not have a college degree and students with at least one parent who had a college degree.

In the completed post-test sample ($N = 386$), we saw a reversal in the attitudes toward engineering. Students with no college-educated parents had the highest scores on attitudes toward engineering, compared to students with at least one college-educated parent. Additionally, the differences in the importance of feedback remained significant, with students without college-educated parents showing the lowest scores. However, the matched sample analysis ($N = 259$) shows none of the between or within group growth was statistically significant.

3.2. Student Open-Ended Responses, Pre- versus Post-Test

Using the two systematically coded student survey narrative questions for the matched sample for Year 2 ($N = 215$) ("What are your views of engineers?" and "What are your plans after high school?") provided a different form of pre/post-test analysis.

Based on coded responses to open-ended questions, the majority of students (79.1%) who entered the EPICS High program expressed a favorable view of engineers at the start of the program (see Table 4). Only three students (1.4%) had a negative view of engineers, and the remaining students (19.5%) had a neutral view of engineers or did not comment. After an academic year of participation in the EPICS High program, there was a decrease in students who had previously provided a negative or neutral view of engineers and engineering (0.5% and 13.1%, respectively) and an increase in the number of students who had a positive view of engineers and engineering (84.4%) (noting the actual numbers of students coded differently between pre/post-test are few in number). Here are two exemplar quotes of students' pre- and post-test answers regarding their views on engineering.

Table 4. Coding of responses on student views of engineers.

	Positive (%)	Negative (%)	Unknown/Neutral (%)	TOTAL (%)
Pre-test	170 (79.1)	3 (1.4)	42 (19.5)	215 (100)
Post-test	184 (86.4)	1 (0.5)	28 (13.1)	213 (100)

Female, Latina, 12th Grade at non-Title I School:

Pre-test answer: "Boring."

Post-test answer: "I think they are amazing at what they do and make it interesting."

Female, Asian, 11th Grade at non-Title I School:

Pre-test answer: "I don't really know much about engineers but I am interested in exploring."

Post-test answer: "What I've learned is that engineers are those that work on certain projects to better the economy. There are engineers who build things and engineers who work with other things such as computers and technology. There are many types of engineers. There are civil and there are aerospace [engineers]. Personally, I'm not sure if I want to be an engineer, but I am trying to explore the field of engineering and what I am most interested in is aerospace engineering. The whole concept of engineering intrigues me and I might consider going into engineering in college."

Based on codes applied to the student narrative responses to open-ended survey questions, most of the students who entered the EPICS High program indicated at pre-test that they planned to attend college after they graduate high school (see Table 5). For example, 175 students (80% of the matched responses) indicated they were planning to go to college after high school. This number remained unchanged in the post-test. This confirms that high school students who enter the EPICS High program already want to go to college prior to entering the program. However, there is evidence that their goals become more refined after finishing the academic year, as demonstrated by this exemplar quote.

Table 5. Plans after high school *.

	College	Military	Unsure
Pre-test	175	9	12
Post-test	175	14	15

Note: * Responses could be coded more than once if more than one answer was provided.

Male, Latino, 12th Grade, Title I School:

Pre-test answer: "I want to pursue a career in some form of engineering while learning about entrepreneurship on my own or through a mentor. I plan to obtain a four-year degree in some form of engineering and later work as an engineer while building up my very own business."

Post-test answer: "After high school, I want to attend the Honors College at ASU and major in Computer Systems Engineering. At the same time, I also want to be involved in my community, working alongside my peers to bring solutions to problems."

3.3. Participant Observation/Interview Data Analysis

Based on inductive systematic coding of the qualitative data, three salient themes, with corresponding subthemes, emerged from the data analysis: (1) Increase in engineering self-efficacy and skills; (2) community embeddedness; and (3) increase in resiliency and positive relationship with failure. Below, we illustrate these findings using exemplar quotes from students.

Theme 1. *Increase in Engineering Self-Efficacy and Skills.*

Subtheme 1a: Confidence and self-efficacy. Throughout site visits, students shared the ways in which participating in EPICS High helped to expand their self-efficacy in engineering. Students discussed how program participation allowed them to view themselves as engineers and gain confidence in their engineering skills. To illustrate, when asked what they gained from participating in EPICS High, this student stated:

> "Getting a chance to learn engineering, seeing it's not as hard as I thought. I can see myself possibly doing engineering. I initially thought you had to be super smart." (female, non-White, non-Title I)

From the notated conversations with students, we learned that program experiences served to both demystify engineering and help students to make connections to engineering, especially for students who may have come into the program indifferent or uninterested in engineering.

Subtheme 1b: Skills and real-world engineering experience. Students spoke of how the class enabled them to gain real-world experience and witness how engineering can be used beyond the classroom. Students spoke of how they liked that the class was hands-on, and that they not only are learning but are applying what they are learning as they engage in the EPICS High design process. They often cited how EPICS High serves as an opportunity for them to learn to "think like engineers" and work to implement solutions. Students expressed that working on their EPICS projects helped to increase their critical thinking and problem-solving skills, while fostering creativity.

> "[EPICS] gives you a little taste of what engineering is like. It's a lot of thinking outside the box and handling different problems based on the scenario—it's really cool." (male, non-White, Title I)

> "EPICS forces you to work on a project that doesn't have a right or wrong answer, leaves more room to be creative." (male, White, non-Title I)

Additionally, students often spoke of how they had to learn to construct a project budget and plan, conduct research, use Gantt charts to track task assignments, and increase their technology literacy to learn new software for their projects.

Subtheme 1c: Collaboration and communication skills. While learning how to do their projects, students were able to increase their teamwork and communication skills. Through conversations with students, we learned that project experiences provided opportunities for them to leverage the synergy of different skills, personalities, and competing goals to benefit from intragroup collaboration and creativity.

> "I learned in engineering that it's hard to work on a project by yourself, you need others to help. This is good—with other people, you get new ideas, and this helps change your project for the better." (male, non-White, Title I)

> "It didn't start off well with my group in the beginning. There were two female strong leaders in our group. Eventually, I gained a new friend. I learned not just about her and about the group, I learned to mature and let everyone contribute." (female, non-White, Title I)

"It [EPICS] taught me things I will need to know in the real world. It teaches you how to work in a team and to step outside of your comfort zone. You have to work with someone else [the stakeholder] who agrees with what we said." (female, non-White, non-Title I)

These exemplar quotes highlight that EPICS High provides a space for students to improve their peer-to-peer collaboration skills as well as communicating with community partners, which also fosters unique student gains.

Theme 2. *Community Embeddedness and Civic Engagement.*

Throughout the program, EPICS High asks its students to continually explore problems in the community that can be solved by the skills they are learning in the classroom and identify those in their community that would benefit from an EPICS project. During site visits, students expressed that they liked that their EPICS projects connected them to a community or issue and cited how the class provided opportunities to not only gain experience working with a client, but exposure to skills and values necessary to design with someone else in mind.

Subtheme 2a: Empathy. The curriculum focuses on human-centered design as a core principle. As students spoke of their experiences, many cited that their projects enabled them to foster greater empathy to better understand the viewpoints, social conditions, and needs of their stakeholders and community partners. Through experiences working on a long-term project with a client, students cited the value of empathy for both their persistence to finish and effectiveness of their projects as well as reflecting on their own learning.

"I think our stakeholder is doing more for us. They're helping us have an understanding of their everyday life. It helps us learn more about people different from ourselves . . . I see how my skills can help, but also how they in turn help us as well." (female, White, non-Title I)

Moreover, students expressed that their projects provided value to themselves, stating that their projects were "more than just a grade" for them. Throughout the informal interviews, students expressed that project experiences provided the opportunity to create something of value for a client.

"EPICS wasn't what I expected. There's so much more of an adjustment period- learning to do projects that aren't just for myself but for the betterment of the community." (male, White, non-Title I)

"In EPICS you have to live up to a nonprofit and not a grade. It is more fun and you hold yourself more accountable because you have to help people." (female, non-White, non-Title I)

For many students, working with a client was a new experience and a paradigm shift from traditional STEM projects that may be abstract and hypothetical. Through doing engineering and working with an actual client, students learned to foster empathy as an engineer and learned to position the user at the center of the design solution.

Subtheme 2b: Impact in the community. Site visits also served as an opportunity to learn the ways in which EPICS High serves as a conduit for students to learn more about engineering (and STEM) through service. Throughout the informal interviews, students often discussed the benefits of having a direct impact on the community. In fact, an overwhelming number of students expressed that they were drawn to the program for the opportunity to impact and help the community.

"As a minor [under 18 years of age], it's hard to impact the community. In STEM, you can have an impact on the community." (female, White, non-Title I)

"EPICS is to serve the community—all of our projects are school-based. The idea of the class is to give and help the community—this was a big push for a lot of us I feel, doing something positive and being a role model to younger students." (male, non-White, Title I)

During discussions with students, we found that they were able to articulate with whom they are working, but also how their stakeholders would benefit from their projects. From the site visits, we see that students' experiences working on projects helped them to increase their awareness that they can play a part in fixing problems in their community as the desire to impact the community taps into students' intrinsic motivation. Ultimately, after working on their projects, students stated how they are able to see how engineering can be used to help the community.

Subtheme 2c: Personal connectedness to projects. EPICS High provided opportunities for students to not only impact their communities but make individual connections to their projects as they impact their communities. Moreover, students expressed that by working with and designing solutions for members of their local community, they were able to identify and connect the ways in which their skills and interests can be used to creatively solve problems.

" ... I've always been passionate about the environment. In class, I was able to tie engineering and combine two of my passions. In engineering, there's not just one job, you can do many things and tie it to your passions—it's one of the great things about engineering." (female, White, non-Title I)

For many, the EPICS High service component connected students to a problem, need, or interest they were passionate about and enjoyed. This served as an additional source of motivation to adhere to and complete project deadlines when teams experienced unintended setbacks.

"Other projects you do for other classes aren't personal—you're just following instructions to get things done; with this class, it's more personal." (female, non-White, Title I)

Additionally, students discussed that while working on their projects, tasks were assigned according to each person's skills, experience, strengths or interests, further enabling students to make personal connections to their projects. Doing a project for an individual purpose helped students to be empowered in their learning and to take more ownership of their learning.

Theme 3. *Increase in Resiliency and Positive Relationship with Failure.*

In the EPICS High curriculum, students are encouraged to approach the design process with a mindset that is open to failure, ambiguity, and feedback.

Subtheme 3a: Importance of feedback. An intended outcome of EPICS High is to impart in students a particular philosophy of engineering, which is to create human-centered solutions that provide value for real people, and this includes asking for feedback from their stakeholders. Many students highlighted that gaining experience working with and getting feedback from a client as not only a major pull to participate in the program, but a major takeaway from the program as well.

" ... getting to see what an engineer actually does. Our teacher has made a point that we are engineers and we're getting feedback and criticism like engineers do—we're getting a lot of real- world experience. How else will I get that kind of experience?" (female, non-White, Title I)

"[EPICS High] prepares us for real human interactions between us and our partners. It's our responsibility [to meet their needs]." (female, White, non-Title I)

Additionally, students shared experiences learning how to compose and execute communications with their client and incorporate stakeholder requests and feedback.

Subtheme 3b: Overcoming challenges. Throughout the conversations with students, they expressed that they learned that failure is a critical part of engineering.

"I was raised to be an honors student ... I always heard that failure was okay from music students—I didn't accept this, and it took me a while to accept the concept of failure ... it's a matter of decoding yourself because it challenges how you're traditionally taught." (female, non-White, Title I)

Students often had to pivot, redesign, iterate or restart a project to ensure their solutions were both feasible and of value to their stakeholders. Many students expressed that EPICS High provided a space for them to comfortably fail in a safe environment and gain confidence in their mistakes to increase their learning potential.

"In other classes, failure is a big deal and you only get one chance and it's more punishing to make those mistakes. You do something wrong and get a bad grade and no opportunity to do it better. It feels like you can't make mistakes in other spaces whereas in engineering, failure is seen as progress." (male, non-White, non-Title I)

"The class is a life teaching class. It gives you the opportunity to have trial and error—not a lot of courses provide this." (male, White, non-Title I)

Furthermore, students often expressed that the opportunity to positively impact their communities kept them motivated to persist with their projects, especially when they encountered setbacks or design difficulties. These examples demonstrate how the program provides opportunities for students to learn to embrace their mistakes, evolve when necessary, and foster resilience to increase their learning potential.

4. Discussion

Prior research on problem-based and service-learning programs has also concluded that they promote valuable development of skills and knowledge at the college level [7,11–15]. These findings are echoed with students from the EPICS High program we tested here, and our results bolster prior smaller studies on EPICS High in particular [16,26]. We found that students who enrolled in EPICS High over a one-year period showed increasingly high positive views of engineers and were more likely to recognize the importance of feedback, multiple perspectives, and social responsibility and were more likely to see themselves as resilient to challenges and improving on existing ideas. Narrative pre- versus post-test responses also showed that they learned more about engineering and honed their career goals during the program.

Furthermore, our analysis by student and school characteristics identified who best benefitted from program participation. First, attitudes toward engineering increased for all students, but significantly for female and Asian students. Second, the importance of feedback and the importance of multiple perspectives were significantly different at baseline across gender, Title I schools, self-identified ethnicity, and across students with different levels of parental education. Female students closed the gap at post-test for these scales; however, differences remained for Title I schools and ethnicity. Finally, at post-test, we observed no differences in the importance of feedback across students with different levels of parental education, though the importance of multiple perspectives still showed significantly higher scores among students with both parents having a college degree.

Scalar data also showed that students who participate in EPICS High overall scored high on engineering-related mindsets and attitudes at the outset of programming, suggesting that these students already had a great interest and regard for engineering upon starting EPICS High programs. The ethnographic/qualitative analysis based on site visits and informal interviews provides a crucial additional set of evidence for interpreting the scalar pre/post-test findings, most especially because—while statistically significant student gains were observed—it was also hard to interpret the impact of change since starting values were surprisingly high. Qualitatively, participants increased their confidence as engineers, learned about engineering processes, used and further developed their problem-solving and creative abilities, increased their collaboration and communication skills, and learned to embrace failure as positive and how to persist when challenges arose. For instance, through gaining exposure to the culture of engineering (e.g., failing, rapid prototyping), students said they are able to decrease their aversion to failure and learn to turn it into opportunity. Furthermore, students recognized gains in understanding the applicability of engineering design to solve problems in their community as well as became civically engaged. These results suggest that EPICS High offers meaningful improvements in student outcomes that should help to meet the larger goals of increasing diversity in engineering/STEM. Thus, quantitatively, the student scores suggest EPICS High participation is not creating much change in the skills domains that could support any students', including URM and female students', pathways into Engineering/STEM. However, the systematic analysis of two years of ethnographic (participant observation and interview) data provides a very different set of insights in this regard and suggests the programs are positively impacting URM and female students in particular, and in ways that are meaningful and could potentially orient them toward STEM.

Based on this, we concluded that we are able to identify small but highly personally meaningful shifts in how students align engineering careers with their own diverse backgrounds, particularly the unique needs of the communities they come from and/or wish to serve. Students also expressed a sense of greater resiliency to challenges; together, these two aspects of the experience were highly meaningful to them in ways that could be predicted as important to the likelihood they will persevere in the engineering/STEM tracks even if they encounter barriers. In this way, there are grounds to conclude the EPICS High programs are serving diverse students well, most especially when project-based and experiential learning focuses them on considering and engaging the needs of local communities. This finding—of subtle but important shifts—also highlights the importance of tracking program impacts not just through scalar changes on attitudes and skills, but also using qualitative/ethnographic approaches to identify changes in what students give meaning to as they consider STEM/engineering careers.

We have several suggestions for the next steps to identify the strengths and weaknesses and track the impacts of project-based learning programs like EPICS High for meeting the goal of orienting more students toward and enhancing the diversity of STEM/engineering degree and career pathways. Given that diverse students came to these programs already positively viewing engineers/engineering, it may be better to focus on early educational (e.g., mathematics) interventions and training [17] and/or having much earlier exposure to STEM problem-based learning to show its applicability as a career [18], such as in middle schools.

There are several limitations of the study we highlight. While the liberal comparison of the total completed sample found more significant differences than the conservative matched samples tests, few of the results were replicated in the matched pre/post-test sample. This could be a power issue, given the low sample size. Power analysis indicates that in order to have 80% power to detect an effect close to those observed in the full sample tests (small to medium effect size: $D = 0.1$–0.3) at *alpha* = 0.05, we would need between 90 to 786 matched pairs. While the power analysis indicates we have a sufficient sample size to detect medium effects (~0.3), given the high scores on the scales at pre-test, we would expect smaller effect sizes (i.e., there may be insufficient room for increases to be observed). On a related point, the scales failed to capture optimum levels of variation in response, that is, to avoid further ceiling effects. Results may have been easier to interpret if Likert response scales had been 7-point rather than 5-point. Additionally, more orthogonal scale items could have helped to reduce correlations between scales, helping to identify more concretely the domains in which students are thinking about engineering and engaging with the program content. More generally, all the pre/post-test scale items are self-reported attitudinal measures. The gold standard would be to validate the scales with behavioral measures, such as pre- and post-test engineering problem-solving assessments that can measure the application of curricular elements that students learn [35]. Further, the study as designed did not include a comparison group of students in the same schools but not enrolled in EPICS. This would have helped to identify more clearly any self-selection into the EPICS High program by students that already hold more positive views of STEM. It could have also helped us to better understand differences between schools and student demographic groups in both baseline and post-test scores.

Author Contributions: Conceptualization, A.R., J.V. and T.G.G.; Data curation, J.H. and T.S.; Formal analysis, J.H., T.S. and R.L.; Funding acquisition, A.R.; Investigation, T.S. and J.V.; Methodology, A.R. and T.G.G; Project administration, A.R. and J.V.; Supervision, A.R.; Validation, A.B., J.V. and T.G.G.; Visualization, A.B.; Writing—original draft, A.R., J.H., T.S. and R.L.; Writing – review & editing, A.R. and A.B.

Funding: This research has been supported by the Cisco corporate advised fund at the Silicon Valley Community Foundation, grant number 2016-154100, Research Project.

Acknowledgments: Tirupalavanam G. Ganesh acknowledges this material is based upon work supported by the National Science Foundation under Grant No. 1744539. Any opinions, findings, and conclusions or recommendations expressed in this material are those of the author(s) and do not necessarily reflect the views of the National Science Foundation. We thank Hope Parker for helping coordinate efforts as well as all the teachers and students who participated.

Conflicts of Interest: The authors declare no conflict of interest.

References

1. The National Academies of Science, Engineering, and Mathematics. Report to Congress. 2016. Available online: http://www.nationalacademies.org/annualreport/Report_to_Congress_2016.pdf (accessed on 5 December 2018).

2. NSF 2017. Supplemental Data. Available online: https://www.nsf.gov/statistics/2017/nsf17310/digest/about-this-report/ (accessed on 5 December 2018).

3. Musu-Gillette, L.; Robinson, J.; McFarland, J.; KewalRamani, A.; Zhang, A.; Wilkinson-Flicker, S. *Status and Trends in the Education of Racial and Ethnic Group*; U.S. Department of Education, National Center for Education Statistics: Washington, DC, USA, 2017. Available online: https://nces.ed.gov/pubs2017/2017051.pdf (accessed on 5 December 2018).

4. National Academy of Sciences; National Academy of Engineering, and Institute of Medicine. *Rising Above the Gathering Storm: Energizing and Employing America for a Brighter Economic Future*; The National Academics Press: Washington, DC, USA, 2007. [CrossRef]

5. National Academy of Sciences; National Academy of Engineering; Institute of Medicine Committee on Underrepresented Groups and the Expansion of the Science and Engineering Workforce Pipeline. *Expanding Underrepresented Minority Participation*; The National Academics Press: Washington, DC, USA, 2011.

6. Engineering Projects in Community Service (EPICS) High. Purdue University, West Lafayette, IN. 2018. Available online: https://engineering.purdue.edu/EPICS/k12 (accessed on 5 December 2018).

7. Coyle, E.J.; Jamieson, L.H.; Oakes, W.C. EPICS: Engineering Projects in Community Service. *Int. J. Eng. Educ.* **2005**, *21*, 139–150.

8. Coyle, E.J.; Jamieson, L.H.; Oakes, W.C.; Bernard, M. Gordon Prize Lecture*: Integrating Engineering Education and Community Service: Themes for the Future of Engineering Education. *J. Eng. Educ.* **2006**, *95*, 7–11. [CrossRef]

9. Zoltowski, C.; Oakes, W.C. Learning by Doing: Reflections of the EPICS Program. *Int. J. Serv. Learn. Eng.* **2014**, *9*, 1–32. [CrossRef]

10. Zoltowski, C.; Cummings, A.; Oakes, W.C.; Immersive Community Engagement Experience. Paper Presented at the Socio-Cultural Elements of Learning through Service Session. 2014 ASEE Conference. Available online: https://www.asee.org/public/conferences/32/papers/10076/view (accessed on 11 December 2018).

11. Bell, S. Project-Based Learning for the 21st Century: Skills for the Future. *Clearing House* **2010**, *83*, 39–43. [CrossRef]

12. Mills, J.E.; Treagust, D.F. Engineering Education—Is Problem-Based or Project-Based Learning the Answer? *Aust. J. Eng. Educ.* **2003**, *3*, 2–16.

13. Newman, J.L.; Dantzler, J.; Coleman, A.N. Science in Action: How Middle School Students are Changing Their World Through STEM Service-Learning Projects. *Theory Pract.* **2015**, *54*, 47–54. [CrossRef]

14. Perrenet, J.C.; Bouhuijs, P.A.J.; Smits, J.G.M.M. The Suitability of Problem-Based Learning for Engineering Education: Theory and Practice. *Teach. High. Educ.* **2000**, *5*, 345–358. [CrossRef]

15. Tsang, E. (Ed.) *Projects that Matter: Concepts and Models for Service Learning in Engineering. American Association for Higher Education's Series on Service Learning in the Disciplines*; Stylus Publishing LLC: Sterling, VA, USA, 2000.

16. Nation, S.; Oakes, W.; Bailey, L.; Heinzen, J. Conversion of Collegiate EPICS to a K-12 Program. In *Frontiers in Education, 2005. FIE'05. Proceedings 35th Annual Conference*; IEEE Publications: Indianapolis, IN, USA, 2005.

17. Wang, X. Why Students Choose STEM Majors: Motivation, High School Learning, and Postsecondary Context of Support. *Am. Educ. Res. J.* **2013**, *50*, 1081–1121. [CrossRef]

18. Bairaktarova, D.; Evangelou, D. Creativity and Science, Technology, Engineering, and Mathematics (STEM) in Early Childhood Education. In *Contemporary Perspectives on Research in Creativity in Early Childhood Education*; Saracho, O., Ed.; Information Age Publishing: Charlotte, NC, USA, 2012; pp. 377–396.

19. Marshall, S.P.; McGee, G.W.M.; McLaren, E.; Veal, C.C. Discovering and Developing Diverse STEM Talent: Enabling Academically Talented Urban Youth to Flourish. *Gifted Child Today* **2011**, *34*, 16–23. [CrossRef]

20. Litzinger, T.; Lattuca, L.R.; Hadgraft, R.; Newstetter, W. Engineering Education and the Development of Expertise. *J. Eng. Educ.* **2001**, *100*, 123–150. [CrossRef]

21. Jacoby, B. *Service-Learning in Higher Education: Concepts and Practices. The Jossey-Bass Higher and Adult Education Series*; Jossey-Bass Publishers: San Francisco, CA, USA, 1996.

22. Billig, S.H. Unpacking What Works in Service-Learning. In *Growing to Greatness*; National Youth Leadership Council: Saint Paul, MN, USA, 2007.

23. Billig, S.H.; Root, S.; Jesse, D. The Relationship Between the Quality Indicators of Service-Learning and Student Outcomes. In *Improving Service-Learning Practice: Research on Models to Enhance Impacts*; Root, S., Callahan, J., Billig, S.H., Eds.; Information Age Publishing: Charlotte, NC, USA, 2012; pp. 97–115.

24. Huff, J.L.; Mostafavi, A.; Abraham, D.M.; Oakes, W.C. Exploration of New Frontiers for Educating Engineers through Local and Global Service-Learning Projects. In *Construction Research Congress 2012: Construction Challenges in a Flat World*; American Society of Civil Engineers: Reston, VA, USA, 2012; pp. 2081–2090.

25. Matusovich, H.M.; Oakes, W.C.; Zoltowski, C.B. Why Women Choose Service-learning: Seeking and Finding Engineering-Related Experiences. *Int. J. Eng. Educ.* **2013**, *29*, 388–402.

26. Kelley, T.; Brenner, D.C.; Pieper, J.T. *PLTW and Epics-High: Curriculum Comparisons to Support Problem Solving in the Context of Engineering Design*; Research in Engineering and Technology Education; National Center for Engineering and Technology Education: Lafayette, IN, USA, 2010.

27. Zoltowski, C.B.; Oakes, W.C.; Cardella, M.E. Students' Ways of Experiencing Human-centered Design. *J. Eng. Educ.* **2012**, *101*, 28–59. [CrossRef]

28. Oakes, W.C.; Dexter, P.; Hunter, J.; Baygents, J.C.; Thompson, M.G. Early Engineering through Service-Learning: Adapting a University Model to High School. In Proceedings of the 119th ASEE Annual Conference and Exposition, San Antonio, TX, USA, 9–13 June 2012; American Society for Engineering Education: Washington, DC, USA, 2012.

29. Kern Entrepreneurial Engineering Network (KEEN). Available online: https://engineeringunleashed.com/mindset-matters/framework.aspx (accessed on 11/12/2018).

30. Brunhaver, S.R.; Bekki, J.M.; Carberry, A.R.; London, J.S.; Mckenna, A. Development of the Engineering Student Entrepreneurial Mindset Assessment (ESEMA). Available online: https://advances.asee.org/development-of-the-engineering-student-entrepreneurial-mindset-assessment-esema/ (accessed on 15 November 2018).

31. London, J.S.; Bekki, J.M.; Brunhaver, S.R.; Carberry, A.R.; Mckenna, A. A Framework for Entrepreneurial Mindsets and Behaviors in Undergraduate Engineering Students: Operationalizing the Kern Family Foundation's "3Cs". Available online: https://advances.asee.org/a-framework-for-entrepreneurial-mindsets-and-behaviors-in-undergraduate-engineering-students-operationalizing-the-kern-family-foundations-3cs/ (accessed on 15 November 2018).

32. United States Department of Education; National Center for Education Statistics. *State Nonfiscal Public Elementary/Secondary Education Survey, 2012–2013*; Common Core of Data (CCD): Washington, DC, USA, 2013.

33. Bernard, H.R. *Research Methods in Anthropology: Qualitative and Quantitative Approaches*; Rowman & Littlefield: Lanham, MD, USA, 2017.

34. Bernard, H.R.; Wutich, A.; Ryan, G.W. *Analyzing Qualitative Data: Systematic Approaches*; SAGE Publications: Thousand Oaks, CA, USA, 2016.

35. National Assessment Governing Board. *Technology and Engineering Literacy Framework for the 2014 National Assessment of Educational Progress*; The U.S. Department of Education: Washington, DC, USA, 2013.

MDPI

St. Alban-Anlage 66

4052 Basel

Switzerland

Tel. +41 61 683 77 34

Fax +41 61 302 89 18

www.mdpi.com

Education Sciences Editorial Office

E-mail: education@mdpi.com

www.mdpi.com/journal/education

www.ingramcontent.com/pod-product-compliance
Lightning Source LLC
Chambersburg PA
CBHW041215220326
41597CB00033BA/5973